Research Report

The Foundations of Operational Resilience—Assessing the Ability to Operate in an Anti-Access/Area Denial (A2/AD) Environment

The Analytical Framework, Lexicon, and Characteristics of the Operational Resilience Analysis Model (ORAM)

Jeff Hagen, Forrest E. Morgan, Jacob L. Heim, Matthew Carroll

RAND Project AIR FORCE

Prepared for the United States Air Force
Approved for public release; distribution unlimited

For more information on this publication, visit www.rand.org/t/RR1265

Library of Congress Cataloging-in-Publication Data is available for this publication.

ISBN: 978-0-8330-9202-1

Published by the RAND Corporation, Santa Monica, Calif.

The RAND Corporation is a research organization that develops solutions to public policy challenges to help make communities throughout the world safer and more secure, healthier and more prosperous. RAND is nonprofit, nonpartisan, and committed to the public interest.

RAND's publications do not necessarily reflect the opinions of its research clients and sponsors.

www.rand.org

Preface

Although much work has been done examining ways to improve airbase resilience, these studies have typically focused on narrow aspects of the problem (such as hardening or runway repair). Almost none have attempted to assess a wider range of improvements to evaluate trade-offs between them and identify which one measure or combination of measures would result in the most resilient force posture, theater-wide. Also, previous studies have neither examined how potential adversaries could tailor their attacks to have the greatest effect in negating U.S. airpower capabilities nor considered what resilience improvements would be needed to defeat those strategies. There is clearly a need to view the issue more holistically, given the strategic implications of growing anti-access/area denial (A2/AD) threats in regions where the United States might have to project airpower to protect its interests or its allies and partners.

This study laid the foundation for such a holistic assessment. We developed a logical framework and lexicon for conducting detailed analyses of Air Force operational resilience in A2/AD environments. We began with an initial survey of several regions (Pacific, Southwest Asia, etc.) to bound the problem and identify a robust set of strategic assumptions and planning requirements. Then, employing a methodology informed by sequential game theory, we identified strategies that an intelligent, adaptive adversary might employ to degrade U.S. combat air capabilities and analyzed how to most resiliently posture U.S. airpower in the face of those attacks. We conducted this analysis in the context of efforts to rebalance the joint force in the Asia-Pacific region.

This report explains the study's purpose, background, objectives, and methodology. It further provides a top-level description of the Operational Resilience Analysis Model (ORAM), which was developed and used to evaluate the potential impacts of enemy attacks on U.S. air forces and compare the potential benefits of alternative resilience improvements. Finally, in an appendix, the report provides a lexicon of resilience-related terms that we hope will become a standard for guiding further research in this area. The results of the complete analysis described in this report are captured in a companion document that, for national security reasons, is not available to the general public.

This research builds on previous work that RAND Project AIR FORCE has conducted examining how to mitigate emerging threats in A2/AD environments:

- Stacie L. Pettyjohn and Alan J. Vick, *The Posture Triangle: A New Framework for U.S. Air Force Global Presence*, RR-402-AF, 2013.
- Alan J. Vick and Jacob L. Heim, *Assessing U.S. Air Force Basing Options in East Asia*, MG-1204-AF, 2013, not available to the general public.
- Brent Thomas, Mahyar A. Amouzegar, Rachel Costello, Robert A. Guffey, Andrew Karode, Christopher Lynch, Kristin F. Lynch, Ken Munson, Chad J. R. Ohlandt, Daniel

M. Romano, Ricardo Sanchez, Robert S. Tripp, and Joseph V. Vesely, *Project AIR FORCE Modeling Capabilities for Support of Combat Operations in Denied Environments*, RR-427-AF, 2015.

The research described here was sponsored by the Director, Air Force Quadrennial Defense Review, Office of the Air Force Assistant Vice Chief of Staff, Headquarters U.S. Air Force (AF/CVAR), and conducted in the Strategy and Doctrine Program of RAND Project AIR FORCE as part of a fiscal year 2013–2014 project "Analytical Support to the U.S. Air Force Quadrennial Defense Review Office."

RAND Project AIR FORCE

RAND Project AIR FORCE (PAF), a division of the RAND Corporation, is the U.S. Air Force's federally funded research and development center for studies and analyses. PAF provides the Air Force with independent analyses of policy alternatives affecting the development, employment, combat readiness, and support of current and future air, space, and cyber forces. Research is conducted in four programs: Force Modernization and Employment; Manpower, Personnel, and Training; Resource Management; and Strategy and Doctrine. The research reported here was prepared under contract FA7014-06-C-0001.

Additional information about PAF is available on our website:

http://www.rand.org/paf/

This report documents work originally shared with the U.S. Air Force in September 2013. The draft report, issued in November 2014, was reviewed by formal peer reviewers and U.S. Air Force subject-matter experts.

Contents

Preface iii
Figures vii
Tables ix
Summary xi
Acknowledgments xvii

1. Introduction 1
   - Background 1
   - Research Purpose and Objectives 4
   - Approach, Strategic Assumptions, and Analytical Methodology 4
   - Organization of This Report 7
2. An Overview of the Modeling Approach 9
   - ORAM and Its Key Features 9
   - ORAM's Modules and Relationships 11
3. Modeling Blue Capabilities 13
   - Aircraft and Beddown 13
   - Basing Locations and Infrastructure 14
   - Airbase Repair and Resilience 23
   - Refueling Orbits 26
   - Missile Defenses 29
4. Modeling Red Capabilities 33
   - Red Knowledge 33
   - Missile Capabilities 34
   - Attack Strategies 35
5. Sample Outputs Using ORAM 37
   - How Model Results Are Calculated 37
   - Baseline Case 40
   - Blue Resilience Improvements 48
   - Red Response to Blue's Improvements 49
6. Potential for Follow-On Work 53
   - Improving the Functionality of ORAM 53
   - Addressing Data Uncertainties with ORAM 54
   - Expanding the Game-Theoretic Analytical Framework 56
   - Broader Uncertainty and Sensitivity Analysis 56
   - Redesign to a Multiresolution Version of ORAM 57

Appendix A. Operational Resilience Lexicon 59

Abbreviations 71

Bibliography 73

Figures

Figure S.1. Operational Resilience Analysis Framework xiii
Figure 1.1. Operational Resilience Analysis Framework 6
Figure 2.1. ORAM Modules and Relationships 11
Figure 3.1. Fitting TBM Aimpoint and Parking Area Data 20
Figure 3.2. Reagan National Airport Imagery 21
Figure 3.3. Reagan National Airport Imagery with Weapon Footprint Diameters Overlaid 22
Figure 3.4. Reagan National Airport Imagery with Weapon Footprints Overlaid 23
Figure 3.5. Illustration of a Runway Cut 24
Figure 5.1. Sortie Capability by Aircraft Type, Prior to Attacks 43
Figure 5.2. Sortie Capability by Location, Prior to Attacks 43
Figure 5.3. Sortie Capability by Aircraft Type, Prior to and After Initial Attack (End of Day 1) 44
Figure 5.4. Sortie Capability by Location, Prior to and After Initial Attack (End of Day 1) 45
Figure 5.5. Operational MOEs over Time 47
Figure 5.6. Operational MOEs with Resilience Improvements 49
Figure 5.7. Operational MOEs with Red's Response to Blue's Resilience Improvements 50
Figure 5.8. Operational Sorties Compared 51

Tables

Table 2.1. Key Resilience Questions Requiring Exploration 10
Table 3.1. Sample of Location-Refueling Orbit Matching 28
Table 5.1. Aircraft Beddown Used in Example Case 40
Table 5.2. Basing Location Details 41
Table 5.3. Baseline Resilience Measures at Each Location 42

Summary

Introduction

In the face of growing anti-access/area denial (A2/AD) challenges, the United States must posture airpower to accomplish missions while under intense and persistent attack. In other words, U.S. air forces must have *operational resilience*: the capacity to withstand attack, adapt, and generate sufficient combat power to achieve campaign objectives in the face of continued, adaptive enemy action.[1] Although several analytic efforts at RAND and elsewhere have studied selected elements of Air Force base resilience, little has been done to determine how potential adversaries could tailor their attacks for greatest effect in negating U.S. airpower capabilities or to assess a wide range of potential resilience improvements to evaluate trade-offs between them and identify which one or combination of measures would result in the most resilient force posture, theater-wide.

Given this gap, the U.S. Air Force asked the RAND Corporation to develop a logical framework for assessing Air Force operational resilience in an A2/AD environment. Specifically, RAND was given four objectives: (1) identify the fundamental strategic assumptions about the emerging geostrategic environment as they relate to operational resilience and emerging A2/AD threats, (2) develop a lexicon of resilience-related terms and definitions that could serve as a standard for use across the analytical community, (3) develop an analytical framework for evaluating the potential impacts of alternative attacks on U.S. air forces, theater-wide, and the potential benefits of a wide range of approaches for improving operational resilience, singularly and in combination, in the face of those attacks, and (4) employ the foregoing assumptions and tools in an exploratory analysis to gain insights into approaches for improving Air Force operational resilience within the context of efforts to rebalance the Joint Force in the Asia-Pacific region.

How We Developed the Framework for Assessing Operational Resilience

We began the research with a broad survey of the geopolitical environment of several regions around the globe to bound the problem and identify a robust set of strategic assumptions. That survey led us to a set of key assumptions that shaped our understanding of A2/AD challenges and the need to improve operational resilience given these challenges:

---

[1] The term *resilience* is defined here as a more specific form of the dictionary definition of "springing back into shape" or "recovering quickly from difficulty." Other similar terms that might be used include *robust*, *agile*, and *adaptive*.

1. U.S. forces may not be postured in a resilient manner, particularly in the Western Pacific, making them vulnerable to attack and degrading their effectiveness.
2. Competition for U.S. defense resources will not allow large expenditures to resolve resilience shortcomings.
3. Cost-effective strategies can and must be found to reposture the force for greater resilience.
4. Improving operational resilience will both improve U.S. combat capabilities and strengthen deterrence and crisis stability.

As this initial survey was under way, we began developing an operational resilience lexicon, which drew heavily on the *Department of Defense Dictionary of Military and Associated Terms* (JP 1-02, 2014), because the terms and definitions provided there have been coordinated with and agreed upon by the military services and other agencies within the defense community. However, a number of concepts central to operational resilience are not addressed in the dictionary, and those that do appear are often defined at too general a level to be analytically useful. Therefore, working in concert with the sponsor, other interested offices within Headquarters Air Force, and offices at Headquarters Pacific Air Forces, we developed additional definitions to describe various resilience approaches and added granularity to existing definitions where needed to make them more useful as metrics for analysis.

To calculate the potential impacts of Red (enemy) attacks and the improvements that might be gained from different combinations of potential resilience investments, we developed the Operational Resilience Analysis Model (ORAM), which measures the complex relationships between enemy attacks, resilience improvements, and combat power. ORAM proved to be an essential tool for our effort, and the bulk of this report is devoted to briefly documenting its parameters and explaining its operations (Chapters Three and Four for Blue and Red model inputs, respectively).[2] Yet, important as it was, simply loading ORAM with Blue (friendly) airpower posture data and running a series of Red missile attacks through it would not have adequately captured the dynamics that can emerge in a long-term contest with an intelligent, adaptive adversary. Therefore, we developed an analytical framework informed by sequential game theory in which to structure our analysis and guide our selection and setup of ORAM case runs. That framework is shown in Figure S.1.

---

[2] This volume is not intended to be a detailed analyst manual for ORAM, however.

Figure S.1. Operational Resilience Analysis Framework

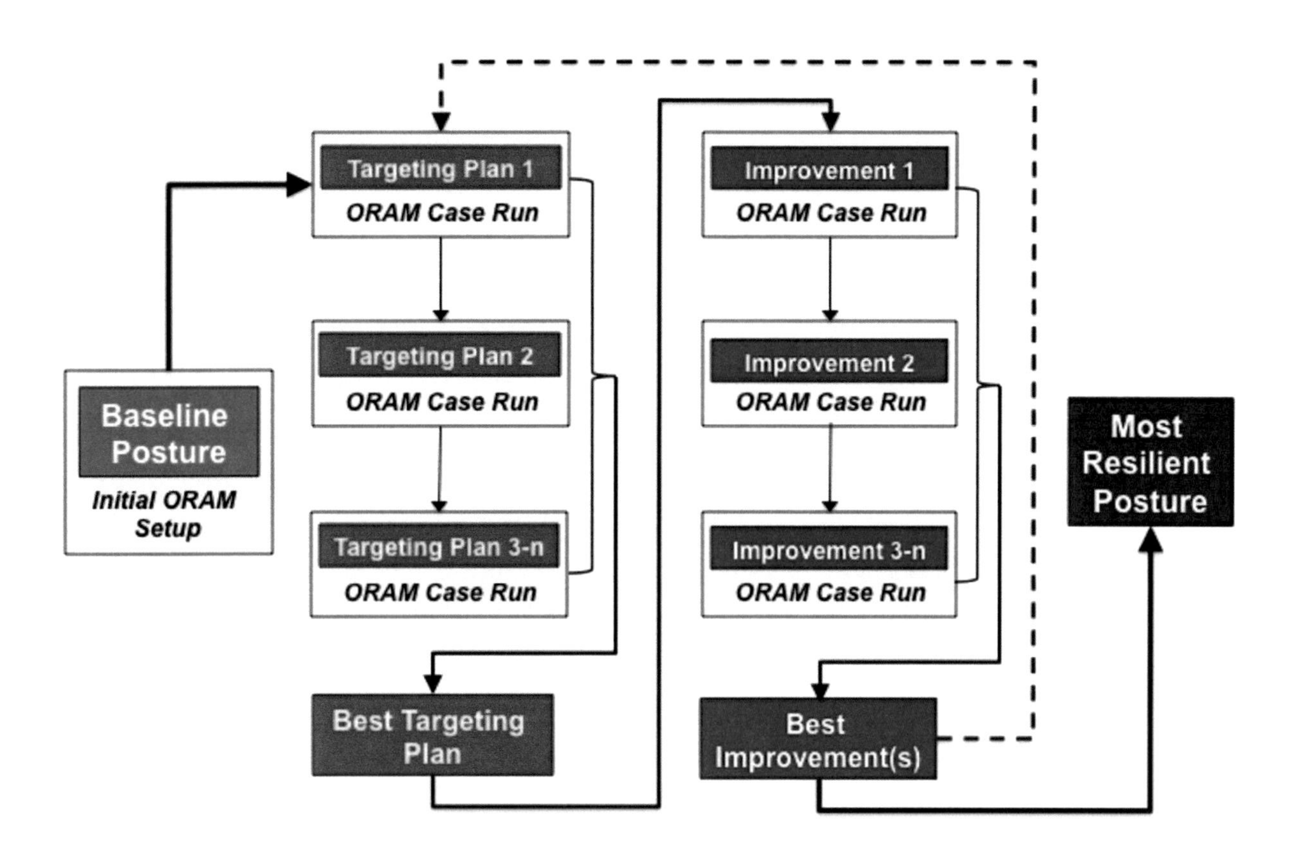

As the figure illustrates, the operational resilience analysis was a multistage process in which we ran multiple cases through ORAM and adjusted each side's strategies based on what the outputs of those runs revealed. We began by populating ORAM with data detailing base infrastructure throughout the theater and the Blue and Red forces available, and we bedded down the force in a baseline posture. Then, we began Turn 1 of the analysis, in which we ran a series of runs examining different Red missile attacks against Blue forces in the baseline posture, with ORAM measuring how effective each targeting plan was in reducing Blue's ability to generate the kinds of sorties it would need to accomplish mission objectives. Running multiple cases revealed which Red targeting plan from those examined would perform best in such a scenario.[3] Knowing that Red is intelligent, we must assume it will recognize that this targeting plan is a very effective way to attack the Blue force and use it in the event of a conflict.

With Red's most effective targeting plan identified, we began Turn 2 of the iterative analysis, experimenting with a series of Blue resilience improvements. Measures such as providing

[3] Here, and elsewhere in the report, when we refer to Red's "best" targeting plan or Blue's "best" resilience improvement(s), we are referring to the most effective option or combination of options from among those we examined (approximately one dozen in this initial effort). This should not be confused with an optimization analysis in which theoretically *all* possible alternatives are evaluated and the one that offers the best trade-off between benefit and cost is identified.

aircraft shelters, adding rapid runway repair capabilities, dispersing aircraft to greater numbers of bases, and moving selected assets farther from the threat were all examined, singularly and in combination, in a series of ORAM case runs. Eventually, ORAM case runs revealed the combination of resilience improvements that best restored Blue combat power—i.e., the one that restored the highest percentage of Blue sorties able to reach the fight—which were, collectively, Blue's best improvements.

However, the analysis did not end at this point, because we had to assume that Red, being intelligent, is also adaptive. Most of the measures Blue could take to make its force more resilient take time to build or otherwise put into effect. During that time, Red can monitor Blue's progress in the effort to strengthen its posture and change its strategy accordingly. Therefore, as the dotted line in Figure S.1 illustrates, we then loaded Blue's best improvements into ORAM and, again, ran a series of case runs against it, exploring how Red could modify its targeting to restore its ability to degrade Blue's combat effectiveness. Once a highly effective revised Red targeting plan emerged, we did another series of ORAM case runs experimenting with additional Blue improvements in efforts to identify how to best defeat the changes in Red strategy. In sum, the analysis was interactive; it was conducted in cycles until the most likely enemy attacks and most needed resilience improvements were clearly evident. Chapter Five provides a simple example that illustrates this interactive process.

Once a set of Blue resilience improvements was identified that was robust against changes in Red targeting strategies, we performed a high-level cost analysis to establish a rough estimate of what that portfolio might cost.[4] Details of the actual analysis and the results we obtained are provided in a companion document that, for national security reasons, is not available to the general public.

This type sequential game should be treated as simply a first step in examining likely strategies for both sides. For example, this approach may not produce the most robust strategy, i.e., the one that produces the most beneficial effect across a range of adversary strategies. It may not produce an optimum strategy for even a single strategy. Thus, while this turn-by-turn approach is a large improvement over simply using doctrine to pick a single strategy for the adversary and planning against it, a full analysis of resilience options would need to expand upon it.

[4] The analysis should continue to also determine Red's most robust strategy, not simply the one that best reduces Blue's sortie generation with a set of resilience-enhancing measures. This would be the analysis analogous to that conducted to find a robust Blue strategy. Together, these would presumably be the strategies that both sides would initially employ. Also interesting to compare would be results between doctrinal strategies and the most robust. The ability of each side to profit by observing the adversary and responding, rather than moving first, would also be important to quantify.

Potential Follow-On Work

Although the work performed in this project was intended to allow both broad and deep analysis of a complex set of resilience issues, there are several areas in which further follow-on work could be done: improving the functionality of ORAM, addressing data uncertainty in the model, expending the interactive analytical framework in which the model is employed, and performing broader uncertainty analysis via increased automation or perhaps a new, multiresolution version of ORAM.

The largest and likely most important of the model **functionality** improvements is including the ability to look at the effect of damage to the logistics and supply chain supporting each airbase. Supporting equipment and personnel, such as ports, road and railways, maintenance hangers, spares, test equipment, munitions, and food, water, and health care for maintainers and pilots, are likely critical to sortie generation, and yet there do not appear to have been recent analyses of their susceptibility to attack or the effects of their loss or damage.[5]

Other improvements that could be made to increase the functionality of the model include (1) modeling Red attacks to interrupt the resupply of fuel to an airbase, (2) more detailed modeling of defenses against cruise missiles, (3) automating refueling orbit locations, and (4) a "wrapper" around the model to automate exploratory analysis.

In the **data uncertainty** area, one key example is the lack of data on the performance of ballistic and cruise missile defenses, which have never been tested against the large raids with countermeasures that are expected in many scenarios. Even cruise missile defense by fighter aircraft, which may be the most straightforward approach, has seen little real-world testing. Other areas with large gaps in data availability include many types of repair times and repair capacity; infrastructure operability at less well-known locations; some types of information on Red capabilities, such as longer-term ballistic and cruise missile operations and warhead availability; and Blue fuel storage and fuel offload infrastructure repairs. Although uncertainty in these and other key variables will never be eliminated, these areas are particularly less informed by real-world data.

The **expanding the game-theoretic framework** used in this research allowed us to explore the interactive nature of the competition between A2/AD developments and operational resilience investments; however, the simplified way in which we applied it introduced a couple of artificialities—the use of notional scenarios and the lack of different Red strategies—that can be avoided in future analyses. Rectifying these deficiencies can enhance the utility of future operational resilience analyses using the interactive framework. Instead of executing generic Red targeting plans in conflicts of fixed duration that Blue attempts to maximize sorties against, researchers should construct baseline scenarios that capture not only baseline estimates of what a potential adversary's campaign plans would resemble, but also a wide range of possible Red

---

[5] Ongoing work at RAND as part of the Combat Operations in Denied Environments (CODE) project is examining these issues in more detail.

targeting plans based on a broad set of excursions around the standard planning scenarios. In conjunction, Blue would be attempting to generate the numbers and types of sorties that achieve a variety of useful real-world objectives (as opposed to simple "more is better" measures of effectiveness). Interactive analysis could proceed from there, setting conflict duration based on possible Red campaign objectives, parameterized to measure the effects on each side's strategies.

As for **exploring a greater portion of the uncertainty space with ORAM**, the interactive analytical framework we employed relied on the analyst to examine and manually adjust Red and Blue strategies at each step. But much of this iterative process could be automated. Automation would enable a wider range of strategic options to be considered and could also be married with an optimization algorithm to find the most cost-effective resilience improvements in the face of an adaptive Red force. This exploratory approach is also quite useful when facing input and strategy uncertainty. If the tool can be expected to behave appropriately with a minimum of human intervention, then computational power can be harnessed to explore very large parametric spaces. This type of analysis is particularly useful to highlight sensitive variables or key areas of uncertainty, as well as to determine the boundaries of useful capabilities. This allows decisionmakers to avoid needless investments and preferentially invest in high-payoff options. One precursor to this approach may be to implement variable resolution modules into ORAM so that broad excursions can be explored with lower-fidelity models and more-detailed results obtained with higher-fidelity algorithms as well.

Some of this work is already underway in various RAND follow-up efforts, primarily in the Strategic and Operational Aspects of Resilience (SOAR), Combat Operations in Denied Environments (CODE), and Shaping Expeditionary Medical Operations in Denied Environments studies. These studies, and others, are examining issues such as logistics issues to support cluster basing; effects of attacks on munitions; fuel supply chain issues, including attacks on ports, pipelines, and railways; investigating attacks on spares and support equipment; and attacks on personnel and the links between personnel availability and operational metrics. For some examples of this ongoing work, see Thomas et al. (2015).

Acknowledgments

This study benefited greatly from the vision of the study sponsor, Lieutenant General Steven Kwast (then Director, Air Force Quadrennial Defense Review, Office of the Air Force Assistant Vice Chief of Staff, Headquarters U.S. Air Force [AF/CVAR], now Commander and President Air University, Maxwell Air Force Base, Alabama). He shaped the definition of operational resilience and, therefore, the resulting analytic framework with his strategic insight on the importance of planning for an adaptive adversary. Thomas "Sam" Szvetecz, our action officer, provided excellent support and shared his expertise as an analyst throughout the project. The authors also benefited from information and insights from Brigadier General Steven Basham (Director of Strategy, Plans, and Programs, Headquarters Pacific Air Forces), who met with the project team to discuss Pacific Air Forces' insights on resilience and its role in maintaining deterrence in the Asia-Pacific region. David Ochmanek (then Deputy Assistant Secretary of Defense for Force Development) took the time to meet with the project team during the project to review the interim results and discuss OSD's concerns about resilience.

At RAND, the authors benefited from the efforts of other members of the project team: Pat Mills (cost analysis), Matt Carroll (airbase data), and Kyle Brady (airbase data). Outside the project team, we wish to thank Bob Tripp's "Combat Operations in a Denied Environment (CODE)" research team, who, working at the same time that this research was under way, shared valuable data they had collected, including alternative beddowns, the capabilities and costs of new shelter designs, and rapid runway repair capabilities. We also wish to thank Jim Chow, Paul Davis, Michael Kennedy, and Pat Mills for their rapid review of this project; Paul Davis, Paul Dreyer, and Brent Thomas for their reviews of this document; and Paula Thornhill for comments and suggestions.

1. Introduction

Background

Being able to project military force into distant theaters is an essential element of U.S. statecraft. Many of the nation's important political and economic interests are located overseas, and the majority of military conflicts in which the United States has been involved have taken place far from the U.S. homeland. To exert influence in distant regions, the United States has maintained capabilities for conducting large-scale movements of military equipment into foreign theaters and posturing forces in ways that communicate to friends and potential adversaries that it can protect its interests.

The U.S. military has resolutely maintained its ability to operate in Europe and East Asia since the end of World War II. While participating in postwar reconstruction efforts, the United States established bases throughout Europe and the Asia-Pacific region, creating a presence that has both played a central role in ensuring the security and stability of those regions and been a source of positive influence in regional politics. During the Cold War, the United States postured forces in these areas to deter expansion by the Soviet Union, People's Republic of China, and Democratic People's Republic of Korea, and U.S. bases in north and southeast Asia supported military operations during the Korean and Vietnam wars.

Since the 1970s and as the Cold War faded in importance, U.S. military forces have also been postured in Southwest Asia, which allowed the United States to prosecute Operation Desert Storm (ODS), Operation Iraqi Freedom (OIF), and Operation Enduring Freedom (OEF). In all these eras, the ability to project power has secured U.S. interests and underwritten the security of U.S. friends and allies in these regions.

Outside of threats by the Soviet Union in Europe, this ability went largely unchallenged. The U.S. military enjoyed decades of access to bases, allowing it to engage in major operations throughout the world. Although some states opposed the basing of U.S. forces in their regions, most of them were unable to effectively threaten those bases or contest the deployment of additional forces into the theater during confrontations or conflicts. However, rapid changes in the international landscape indicate that U.S. enjoyment of broad military access to important regions may be coming to an end. The growing availability to potential adversaries of increasingly potent anti-access/area denial (A2/AD) capabilities pose a serious threat to the ability to deploy U.S. forces into theaters overseas.

Impetus for the A2/AD Threat

ODS was a game-changing military action. With the entire world watching on television, coalition forces defeated what was then the world's fourth-largest military with surprisingly little effort. Large-scale air attacks led by the United States, which Iraqi defenses were unable to stop, attrited large numbers of Iraqi ground forces and other targets from sanctuary bases. World leaders learned from this conflict that there was a need to prepare for future warfare that would take place under what some military strategists began to describe as "high-technology conditions."[1]

The freedom with which coalition members' aircraft were able to deploy to and operate within the conflict area was a critical enabling factor in the coalition's air attack. The United States and other coalition members were able deploy their forces to bases and waters surrounding Iraq and operate there in conditions of relative immunity from enemy attack. Launching from bases in Saudi Arabia and from carriers in the Red Sea and Persian Gulf, aircraft were able to drop 85,000 tons of bombs over the campaign with historically low loss rates (Schneider, 2004). The near-constant presence of airpower over the battlefield made it difficult for Iraqi force elements to move without exposing themselves to destruction. Iraq was unable to defend itself, and resistance quickly collapsed in the 100-hour coalition ground operation that followed the 39-day bombardment.

Other potential opponents learned from this startling example. To mitigate the U.S. airpower advantage in future conflicts, many countries began to invest in means of keeping enemy fighters and bombers from enjoying the same freedom of movement. The concepts of operation that emerged from this desire came to be collectively known as A2/AD.

Manifestation of the A2/AD Threat to Airpower

The growing availability of increasingly potent A2/AD capabilities to potential adversaries poses a serious threat to the ability to deploy and employ U.S. forces into some theaters overseas. As a concept, A2/AD encompasses two separate but related categories of investment. The first, anti-access (A2), refers to a fundamental shift in military investment strategy. A2 measures are designed to render opposing forces unable to make an initial entry into a theater of operations. The People's Liberation Army's development of its arsenal of theater ballistic and long-range cruise missiles is a recent, well-publicized example of an A2 measure. By maintaining a large store of accurate, very long-range missiles, the People's Liberation Army has significantly raised the potential cost of any U.S. military operation in the Asia-Pacific region (Office of the Secretary of Defense, 2013; Krepinevich, Watts, and Work, 2003).

Area denial (AD) refers to any attempt to limit the freedom of movement of attacking forces and protect important targets by putting attackers at unacceptable levels of risk. This generally involves upgrades to ground-based air defenses and air interceptors. Many of America's

---

[1] This refers to Chinese military theorists. People's Liberation Army doctrine had, since the 1980s, made reference to "local, limited war under modern conditions," reflecting the prevailing Chinese belief that major nuclear war was less likely in the modern era. It was not until 1993, though, under Jiang Zemin, that the term *high-technology* replaced *modern*. Several years later, the qualifier was changed again from *high technology* to *informationalized* to reflect the growing role of cyberspace within high-technology conflict. See Cliff, 2011.

potential regional adversaries have purchased or are developing advanced AD capabilities, such as the Russian-made S-300/S-400 surface-to-air missile (SAM) system and China's J-20 fighter, which are far more capable of threatening next-generation U.S. aircraft than earlier designs. In addition to possessing vastly improved capabilities to track and engage targets, many of these systems are highly mobile and far more survivable than any of the air defenses coalition forces faced in Iraq.

Note that many threats can provide both A2 and AD capabilities. The cruise and ballistic missiles mentioned above, for example, not only prevent forces from deploying to bases and entering the theater, but may also limit the sorties flown from those bases and hence provide an AD effect as well. In this work. we primarily focus on A2/AD threats to airpower, but naval and ground forces can be greatly affected as well.

The Need to Study Operational Resilience Holistically

In the face of the A2/AD challenges posed worldwide, the United States must posture its airpower to accomplish its missions while under concerted attack. This suggests the need for an academically rigorous treatment of resilience at the theater level. Much work has been done in this area, initially during the Cold War (see Emerson, 1982, for a RAND example) and then more recently as improved A2/AD capabilities have again come to the forefront (Stillion and Orletsky [1999] and Krepinevich, Watts, and Work [2003] are two key unclassified works). Previous resilience studies tended to focus on base-level survivability, examining topics such as hardening, dispersion, and rapid runway repair in relative isolation.[2] While informative, narrowly focused studies of base resilience are insufficient, because the piecemeal results typically presented to decisionmakers from these analyses focused on individual factors, at best giving only partial answers and at worst providing misleading recommendations. What is needed is a more holistic study of *operational resilience*.

Operational resilience is defined here as *the capacity of a force to withstand attack, adapt, and generate sufficient combat power to achieve campaign objectives in the face of continued, adaptive enemy action.*[3] It is a broad concept that goes beyond the direct engagement of enemy forces. In addition to fielding sufficient troops and equipment to achieve U.S. military objectives,

---

[2] There are several reasons for this. Most of them relate to analytical challenges stemming from how diverse and interconnected the resilience problem can be. For example, a detailed model that captures runway cratering does not provide much insight into campaign-level effects, unless its inputs reflect the presence of missile defenses and its outputs feed into a model of runway repair capability. Similarly, issues as varied as underground fuel hydrant offload rate, aircraft minimum takeoff and landing distances, concrete set time, and ballistic missile reload rates all come into play, challenging both the expertise of the analysts and the ability of the modeling community to capture all of the interrelationships in an analytically tractable way.

[3] This definition was developed in close cooperation with the project sponsor, Director, Air Force Quadrennial Defense Review, Office of the Air Force Assistant Vice Chief of Staff, Headquarters U.S. Air Force (AF/CVAR). It emphasizes two points: first, that U.S. forces must be sufficiently resilient under attack to generate the combat power needed to achieve campaign objectives; and second, that the adversary is intelligent and will adapt to U.S. forces' efforts to improve their resilience.

an operationally resilient force must be able to maintain a basing and support structure capable of sustaining combat operations over the length of a campaign, all in the face of enemy efforts to defeat those activities. U.S. forces and the logistical network supporting them must be capable of operating without significant interruption and adapting to any attempt by opposing forces to compromise them. This is a complex problem with interdependent relationships between many diverse factors.[4] Nevertheless, it is one that must be addressed, given growing A2/AD threats in the emerging strategic environment. This report offers an analytical framework for examining the interplay of the many elements that affect operational resilience.

Research Purpose and Objectives

Given the challenges explained above, the project sponsor, Director, Air Force Quadrennial Defense Review, Office of the Air Force Assistant Vice Chief of Staff, Headquarters U.S. Air Force (AF/CVAR), asked RAND to develop a logical framework for assessing Air Force operational resilience in an A2/AD environment. Specifically, the Air Force asked RAND to accomplish four objectives:

1. Identify the fundamental strategic assumptions about the emerging geostrategic environment as they relate to operational resilience and emerging A2/AD threats
2. Develop a lexicon of resilience-related terms and definitions that could serve as a standard for use across the analytical community.
3. Develop an analytical framework for evaluating the potential impacts of alternative Red (enemy) attacks on U.S. air forces, theater-wide, and the potential benefits of a wide range of approaches for improving operational resilience, singularly and in combination, in the face of those attacks.
4. Employ the foregoing assumptions and tools in an exploratory analysis to gain insights into approaches for improving Air Force operational resilience within the context of efforts to rebalance the joint force in the Asia-Pacific region.

Approach, Strategic Assumptions, and Analytical Methodology

The nature of the analytical tasks assigned in this study indicated the need for a top-down, strategic approach. Therefore, we began the research with a broad survey of the geopolitical environment of several regions around the globe to bound the problem and identify a robust set of strategic assumptions. That survey led us to a set of key assumptions that shaped our understanding of A2/AD challenges around the world and the need to improve operational

---

[4] Some analyses of A2 measures incorporate the political and economic tools that can be leveraged to deny an opposing force entrance to a region. In the case of China, the central role that the country plays in the economy of East Asia could be used against a U.S. expeditionary force. If China were to pressure America's regional allies to restrict or deny access to U.S. forces, a number of bases central to U.S. military strategy could be kept from operating. This report touches on the effects of these types of nonmilitary factors on a force posture's resilience, but a full analysis of the interplay between operational resilience and the politics of the Asian Pacific region would require a separate study.

resilience in the face of these challenges. The most fundamental of those assumptions can be spelled out in four tenets:

1. U.S. forces may not be postured in a resilient manner, particularly in the Western Pacific, making them vulnerable to attack and degrading their effectiveness.
2. Competition for U.S. defense resources will not allow large expenditures to resolve resilience shortcomings.
3. Cost-effective strategies can and must be found to reposture the force for greater resilience.
4. Improving operational resilience will both improve U.S. combat capabilities and strengthen deterrence and crisis stability.

As this initial survey was under way, we began developing the operational resilience lexicon provided in the appendix of this report. To develop this lexicon, we drew heavily on the *Department of Defense Dictionary of Military and Associated Terms*, because the terms and definitions provided there have been coordinated with and agreed upon by the military services and other agencies within the defense community (JP 1-02, 2014). However, the advantage of the dictionary's broad acceptance was offset by the general nature of the definitions it provides. A number of concepts central to operational resilience are not addressed in the dictionary, and those that do appear are often defined at too general a level to be analytically useful. Therefore, working in concert with the sponsor, other interested offices within Headquarters Air Force, and offices at Headquarters Pacific Air Forces, we developed additional definitions to describe various resilience approaches and added granularity to existing definitions where needed to make them more useful as metrics for analysis.

To calculate the potential impacts of Red attacks and the improvements that might be gained from different combinations of potential resilience investments, we developed the Operational Resilience Analysis Model (ORAM). This model provided a means to measure the complex relationships between enemy attacks, resilience improvements, and combat power. ORAM proved to be an essential tool for our effort, and the bulk of this report is devoted to documenting its parameters and explaining its operations. Yet, important as it was, loading ORAM with Blue (friendly) airpower posture data and running a series of Red missile attacks through it would not have adequately captured the dynamics that can emerge in a long-term contest with an intelligent, adaptive adversary. Therefore, we developed an analytical framework informed by sequential game theory in which to structure our analysis and guide our selection and setup of ORAM case runs. That framework is shown in Figure 1.1.

Figure 1.1. Operational Resilience Analysis Framework

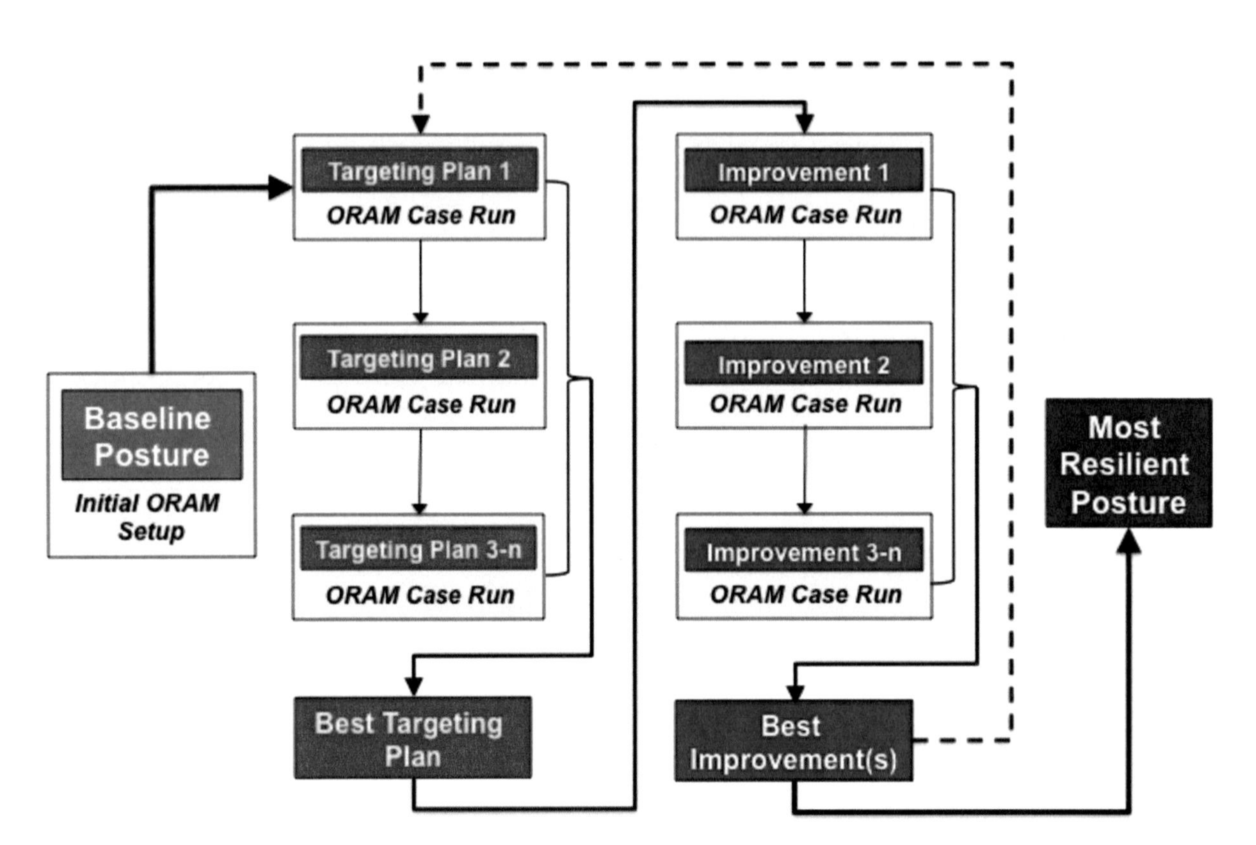

As the figure illustrates, our operational resilience analysis was a multistage process in which we ran multiple cases through ORAM and adjusted each side's strategies based on what the outputs of those runs revealed. We began by populating ORAM with data detailing base infrastructure throughout the theater and the Blue and Red forces available, and we bedded down the force in a baseline posture. Then, we began Turn 1 of the analysis, in which we ran a series of case runs depicting about a dozen different Red missile attacks against Blue forces in the baseline posture, with ORAM measuring the effectiveness of each targeting plan in reducing Blue's ability to generate the kinds of sorties it would need to accomplish mission objectives. Running multiple cases revealed Red's most effective targeting plan, from among those examined, in such a scenario. Knowing that Red is intelligent, we must assume that it will recognize that this targeting plan is a highly effective way to attack the Blue force and use it in the event of a conflict. Here, and elsewhere in the report, when we refer to Red's "best" targeting plan or Blue's "best" resilience improvement(s), we are referring to the option or combination of options from among those we examined that produced the lowest number of Blue sorties. This should not be confused with an optimization analysis in which theoretically *all* possible alternatives are evaluated and the one that offers the largest benefit for the lowest cost is identified.

With Red's targeting plan identified, we began Turn 2 of the iterative analysis, experimenting with a series of Blue resilience improvements. Measures such as providing aircraft shelters, adding rapid runway repair capabilities, dispersing aircraft to greater numbers of bases, and moving selected assets farther from the threat were all examined, singularly and in combination, in a series of ORAM case runs. Eventually, ORAM case runs revealed the combination of resilience improvements we examined that best restored Blue combat power—i.e., the one that restored the highest percentage of Blue sorties able to reach the fight—which were, collectively, Blue's best improvements.

However, the analysis did not end at this point, because we had to assume that Red, being intelligent, is also adaptive. Most of the measures Blue could take to make its force more resilient take time to build or otherwise put into effect. During that time, Red can monitor Blue's progress in the effort to strengthen its posture and change its strategy accordingly. Therefore, as the dotted line in Figure 1.1 illustrates, we then loaded Blue's best improvements into ORAM and, again, ran a series of case runs against it exploring how Red could modify its targeting to restore its ability to degrade Blue's combat effectiveness. Once the best revised Red targeting plan emerged, we did another series of ORAM case runs experimenting with additional Blue improvements in efforts to identify how to best defeat the changes in Red strategy. In sum, the analysis was interactive; it was conducted in cycles until the most likely enemy attacks and most needed resilience improvements were clearly evident.

Once a set of Blue resilience improvements was identified that was robust against changes in Red targeting strategies, we performed a high-level cost analysis to establish a rough estimate of what that portfolio might cost. Details of the actual analysis and the results we obtained are provided in the companion document.

Note that this type of sequential game should be treated as simply a first step in examining likely strategies for both sides. For example, this approach may not produce the most robust strategy, i.e., one that produces the most beneficial effect across a range of adversary strategies. It may not produce an optimum strategy for even a single strategy. Thus, while this turn-by-turn approach is a large improvement over simply using doctrine to pick a single strategy for the adversary and planning against it, a full analysis of resilience options would need to expand upon it.

Organization of This Report

The remainder of this report is devoted to briefly documenting ORAM and the operational resilience lexicon. Chapter Two provides a narrative overview of the modeling approach used to develop ORAM. Chapter Three covers in more detail the Blue force inputs that inform the model. It describes the decisions the user must make and explains how ORAM handles such factors as aircraft selection and beddown, basing locations and infrastructure, airbase repair, the placement of refueling orbits, and missile defenses. Chapter Four explains how ORAM models

Red capabilities. It describes decisions the user must make about missile stores, weapon types, and the quality of Red intelligence, and it explains how ORAM models those elements. The chapter also explains how the user loads and executes each Red attack strategy. Chapter Five explains how model results are calculated and illustrates ORAM's outputs with a simple, notional baseline case. The example demonstrates how to use the model in the interactive analysis described above, showing how adding resilience improvements can increase Blue sortie generation rates and how Red would respond by modifying its firing plan to mitigate those Blue improvements. Chapter Six provides a summary of some options for future follow-on work to improve ORAM and use it more effectively in future analyses. An appendix presents the operational resilience lexicon.

2. An Overview of the Modeling Approach

In this chapter, we provide an overview of the modeling approach—ORAM—used in this effort, starting with what ORAM is and then looking at its modules and relationships. This and the following chapters are not intended to serve as a detailed analyst manual for ORAM, but instead to provide an overview of ORAM's strengths and weaknesses.

ORAM and Its Key Features

To facilitate a quantitative analysis of the theater-wide interactions between an assortment of potential adversary attacks and a range of operational resilience investment choices, a portion of the study effort was focused on developing a new modeling tool—ORAM. The top-level objectives of this modeling effort were to (1) make the inputs and formulas readily accessible to the user, (2) allow Red attack and Blue resilience strategies to interact, and (3) provide operationally relevant output metrics. This chapter provides an overview of how we approached the modeling effort.

To minimize the resources required and take advantage of previous work in many relevant subject areas, we integrated several previously developed tools to create a single unified tool. To accomplish this, we used some models, such as for missile defenses, as they were originally written, whereas we simplified others, such as those for runway damage, to better fit with the campaign-level perspective desired, especially in terms of data-gathering needs. The main integration effort was devoted to the interconnections required, many of which are discussed in this report. In addition, we provided a user-specified time step, to allow for hour-by-hour, day-by-day, or even week-by-week analysis if desired and as necessary for a particular study. The algorithms used throughout are deterministic calculations, so multiple repetitions are not required, as they would be with a Monte Carlo model. This allows for a faster run time and more straightforward input values (since probability distributions are not needed), although, of course, more advanced statistics are not available on the outputs, since only a mean value is being calculated.

One of the challenges with deterministic models is dealing with the uncertainty inherent to many of the input variables. One approach commonly used is termed "exploratory analysis," whereby outputs with a range of likely input values for a broad set of variables are examined. Typically, many of these variables or least portions of their input spaces can be found to be less important to the outputs, and resources can be focused on the variables that have the most leverage. Modern computing power and visualization tools allows millions of cases to be examined fairly quickly, and the exploration of this input space often leads to some of the more important conclusions of an analysis. Although we do not discuss the findings of such a broad uncertainty exploration here, the companion volume does highlight much of our parametric analysis.

The features included in ORAM were determined based on their relevance to several key airbase resilience questions, which can be grouped into categories, as shown in Table 2.1.

Table 2.1. Key Resilience Questions Requiring Exploration

Category	Questions
Dispersal	• Is it more cost-effective to operate from many locations with a small presence at each, or a few hardened locations protected by concentrated defenses?
Distance	• Is operating from outside threat ranges more efficient than operating from closer, but more threatened, areas? • How does distance interact with dispersal, hardening, and the need for air defenses?
Hardening	• How much hardening of airbase infrastructure, such as aircraft shelters or fuel, is cost-effective? • At what locations should this be hardening be done? • Should hardening be done to support all types of aircraft, or only specific ones such as fifth-generation fighters or refueling tankers? • Is hardening particularly useful if used in conjunction with dispersal or air defenses?
Capacity Increases	• Should the number of possible basing locations be increased? • What is the gain in resilience if additional parking areas or runways were built at existing areas? • Would additional air defenses provide a cost-effective benefit?
Repair Capability	• Is runway repair capacity well matched with Red capabilities for attack, both in terms of timing and in total numbers? • Does sufficient capacity and capability exist to repair other airbase infrastructure such as fuel storage and offload?
Air Defenses	• How many defenses and of what type are the most cost-effective? • What types are the most important to minimize the number of leakers? • How survivable are the defenses themselves? • Does the presence of defenses change Red's preferred strategy?

Obviously, not every study will be interested in all of these questions; however, analysts attempting to provide decisionmakers with recommendations on preferred resilience options must at least consider most of these interrelated issues. In fact, the primary utility of ORAM lies in the breadth of factors that it takes into account. By giving high-level treatment to many different aspects of operational resilience, the model helps to identify those that will be critical in determining the outcome of a conflict against a capable Red force. Some of these aspects of resilience have been studied individually at a higher resolution, but this and future modeling efforts are necessary to place them in the larger strategic context. The model was built with the inputs, algorithms, and interactions necessary to address these topics.[1]

[1] Although the ORAM tool itself does not directly address the cost of various resilience options, as discussed in later chapters of this report, a costing framework has also been developed to accompany the effectiveness evaluations performed by ORAM.

ORAM's Modules and Relationships

ORAM is coded in the Microsoft Excel environment with inputs and algorithms as cell formulas for easy visibility and data plotting. A Visual Basic macro is used to advance the model each time step and reset the conflict to the beginning. Each module of the tool is contained in a single worksheet, with interrelationships cross-linked between. The modules are organized roughly in terms of how an actual deployment and campaign would progress over time in the structure plotted in Figure 2.1. Note that this figure depicts the calculations that occur at each time step. In other words, if the model is set to run a 14-day campaign in 1-day increments, it will progress through the full range of computations shown in the Figure 2.1 for each of the 14 days, generating results in terms of sorties flown in support of operational missions each day and cumulatively over the campaign. Similarly, the user can specify shorter time steps, such as 12 or eight hours, if more frequent actions are desired, such as Red attacks or Blue aircraft deployments.

Figure 2.1. ORAM Modules and Relationships

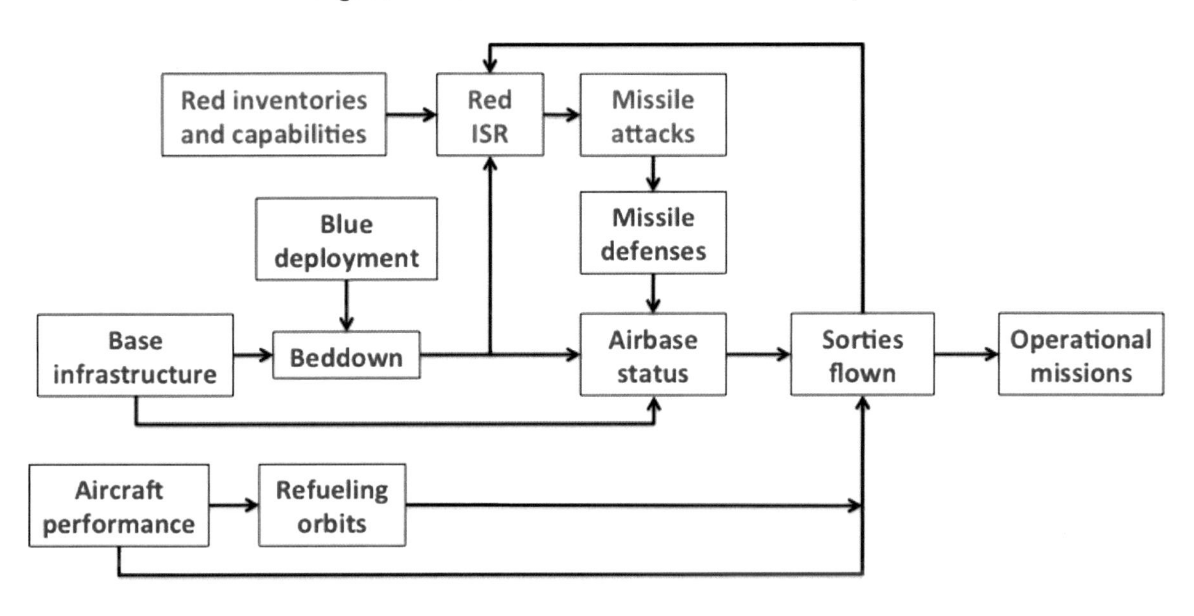

NOTE: ISR = intelligence, surveillance, and reconnaissance.

To use ORAM, analysts first load it with data on the numbers, types, and characteristics of Blue aircraft to be employed in the campaign. The model must also be populated with data describing bases in the theater in which the campaign will occur, the infrastructure available at those locations, and each one's range to the area of operations where the notional conflict will take place. With this information loaded, the user specifies what aircraft will deploy to each location each day of the campaign, depicted in Figure 2.1 as the Blue "Beddown," and designates the locations of refueling orbits—that is, where aircraft will rendezvous with orbiting tankers to get the fuel they need to fly to the area of operations, perform their missions, and return to base.

Finally, the user must designate what missile defenses are available and what locations they will protect.

With Blue inputs complete, the user then enters data describing Red capabilities and how they will be employed. Most of this information revolves around specifying how many missiles of each type Red has, describing their attributes, and developing a plan designating how many of each type will be fired at what targets (locations and aim points) at each time increment. However, a key element for guiding the Red attack strategy in an actual conflict would be access to intelligence about Blue's ongoing operation and how much damage Blue is suffering from Red attacks. Red commanders will not have perfect knowledge about this information. Rather, it will likely be incomplete, partially incorrect, and delayed. ORAM simulates these effects in the module designated "Red ISR" in Figure 2.1, which allows users to specify levels of error in Red targeting and how long, in terms of time steps, it takes Red to see changes in Blue force posture and capability.

Once all Blue and Red entries are loaded, ORAM calculates the results of Red attacks on Blue, time step by time step, over the designated campaign. When each time increment is advanced, it updates the number and types of Blue aircraft, air and missile defense assets, and airbase infrastructure remaining at each location, and also the number and types of Red missiles remaining. Then, the model executes the Red missile attack, based on the firing and allocation logic explained above, and calculates the effects on Blue airbases, aircraft carriers, and the aircraft based there (shown in Figure 2.1 as "Airbase status"). Once the number of available Blue aircraft is determined at each time step, ORAM takes the aircraft performance data and tanker availability (at refueling orbits) into consideration and calculates the key outputs—how many sorties are flown in that time increment and what operational missions each of those sorties support.

In the next two chapters, we will explain how Blue and Red inputs are gathered and modeled in more detail. Then, we will describe how the model performs its time step calculations and provide a few sample results from a simple, notional baseline case in Chapter Five.

3. Modeling Blue Capabilities

As described in Chapter Two, there are many Blue capabilities that need to be modeled in ORAM. Here, we describe the inputs for the various Blue modules shown in Figure 2.1. In cases where specific input values are mentioned below, it is important to highlight that, while these values are currently set in the unclassified version of this tool, they may not be the most accurate for a given scenario or basing location. Obviously, users of any tool should thoroughly examine all of the inputs for accuracy and applicability to the problem under study and engage with subject-matter experts as necessary. We have also tried to highlight cases where great uncertainty is likely to be found—these are obviously inputs that should be explored with parametric analysis whenever possible.

Aircraft and Beddown

The first set of model inputs describes the various aircraft types of interest and their characteristics, primarily the demand for fuel. Twenty different aircraft types are allowed in ORAM (although one type is reserved for tankers), and a key categorization is "small" or "large."[1] This distinction sets the minimum allowable runway length, the submunition spacing needed by attacking ballistic and cruise missiles, and the availability of hardened shelters. In addition, each different aircraft type has inputs for onboard fuel capacity (in pounds), fuel burn (pounds per nautical mile), and the loiter time per sortie to set range and ground and airborne fueling needs.[2] There is also a set of mission allocation inputs that guide the use of operational output measures. So, for example, the "F-22" aircraft type could be set to 100 percent defensive counterair (DCA) mission allocation, and the "F-35" type could be set to 50 percent DCA and 50 percent strike allocation. Thus, the sortie counts of these two types would be used to calculate the supply of DCA orbits, while only the latter would be used to count strike packages flown per day. These mission allocations affect only the output measures in the model—total sorties flown and how many are flown in support of each operational mission. There is no attempt in ORAM to calculate the effectiveness of these sorties on their missions.

The second set of aircraft-specific inputs is the deployments to each location by day. Typically known as the time-phased force and deployment data (TPFDD), these inputs are a series of 20 tables, one for each aircraft type, with the locations as rows and the time steps (up to

---

[1] Ten of each size category can be utilized.

[2] The unclassified version of the model currently uses open-source information for the fuel capacity and fuel burn inputs, primarily from *Jane's All the World's Aircraft* and a DoD-supplied tool known as the Portable Flight Planning Software.

100 of them) as columns.[3] With these tables, the analyst can specify when and how many planes of which types arrive at (or depart) each basing location. In addition, taking advantage of the flexibility of the spreadsheet environment, these entries can also be formulas, allowing the analyst to set various conditions for aircraft movements. So for example, rather than simply having 24 aircraft show up at Base X, the user can input if-then-else logic to deploy them to Base X if Base X is undamaged, or otherwise deploy them to Base Y. Similarly, aircraft could be relocated from damaged bases with low sortie rates to alternative locations that are undamaged.

Basing Locations and Infrastructure

With the aircraft types and capabilities specified, the next set of user inputs revolves around the airbases they will be flying from. The model provides for both sea and land airbases, with up to 100 allowed in a given scenario. The sea bases require only a user-specified location and runway repair times and capacity. Other information, such as runway and parking size, fuel capacity, and offload and sortie rates, are computed internally and based on the capabilities of current U.S. *Nimitz*-class aircraft carriers or landing helicopter assault (LHA)/landing helicopter deck (LHD) amphibious assault carriers. If desired, users could easily modify these capabilities to accommodate other types of ships. Land bases in the model are specified by using real-world names, which are then used to look up all the relevant infrastructure information. The latitude and longitude of each airbase and the loiter time specified for the various aircraft types are used to calculate sortie rates from each location, for both small and large aircraft, and flight times for fuel demands. Specific sortie rates for each location and aircraft type are calculated using a curve fit to outputs from a more detailed model that incorporates such details as turn time, crew ratio, and duration of conflict. Sortie rate differences between the small and large types are mainly the result of longer refuel and rearm times and slower cruise speeds for the latter, although this is somewhat made up for with fewer refuelings. There are also checks to prevent excessively long sorties from being flown that would violate crew duty day restrictions: 12 hours for single-seat aircraft, and 16 hours for those with dual controls.[4] Sortie rates also drop by a user-specified percentage after day D+6 to model the effect of a drop-off in sorties after an initial surge. In addition, the user is provided with an estimate of the number of refueling orbits that will be needed to support each location based on the fuel demands of the aircraft.

For the initial set of cases with the model, we populated the lookup tables for approximately 50 bases throughout the Asia-Pacific region with openly available data. To determine data to be used as inputs to the model, we primarily used the Automated Air Facilities Intelligence File (AAFIF) database; however, in many instances, the AAFIF had incomplete data or data of

[3] The 100 limit on time steps is currently set only by how many formulas have been filled in on the spreadsheet. This could be expanded, in most cases, by simply "dragging" the formulas to additional columns, but there would be a price to pay in run-time due to the additional calculations being performed.

[4] Crew daily duty day and longer-term flight time restrictions are taken from Air Force Instruction 11-202, 2010.

questionable accuracy. To fill in the gaps, we used a number of different techniques, including imagery analysis, an examination of open-source industrial information, and a review of published academic work on airbase logistics. The following subsections outline our techniques and findings for this data-gathering effort, although additional data are available at higher classification levels.

Fuel Storage

The AAFIF provides information on the number and size of fuel tanks present and data entries for tank type, fuel type, and dispensing method for many of our airbases of interest. Where data were lacking, we performed measurements of aboveground fuel tanks by visual inspection. Using Google Earth tools and imagery, we took measurements of fuel storage tanks in the length and width dimensions (since Google Earth's display is a top-down view) and then attempted to estimate the tank's height. Where Google Maps' "Street View" option was available, as it was for several airbases in Japan, it was possible to determine those height measurements directly. Where Street View was not available, it was possible in some instances to estimate the height of a tank by measuring the length of the shadow it cast on the ground. With knowledge of the shadow's length, as well as the local time and the longitude of the airbase's location, we were able to establish the fuel tank's height by triangulation. One shortcoming with this approach is determining the type of fuel being stored in each tank. In some cases, notes were available in the AAFIF database about which tanks were used for which fuel type, but in other cases this approach could result in an overestimation of aircraft fuel capacity.[5]

Where there were gaps in the AAFIF data and measuring fuel tank height was impractical, we attempted to determine the dimensions of fuel tanks with manufacturer data. We compared measurements of the two dimensions available in Google Earth's overhead view to those provided by companies involved in fuel storage tank construction, such as T. Y. Lin International. By surveying the available open-source information across tank manufacturers, we were able to determine average values for tank dimensions and volumes.

To determine fuel storage capacity in an airbase's underground storage, we were forced to rely heavily on AAFIF data. Tanks described as either "partially buried" or "covered" were treated as aboveground tanks if they could be seen on Google Earth. Fully underground tanks, described by AAFIF as tanks with "top at or below natural grade," were assumed to be present or absent on the strength of AAFIF's numbers alone. It is possible that this resulted in some underestimation of the base's total fuel capacity, because any fuel tanks installed after the 2006 final update of AAFIF's data would not have been counted. In a few cases, we used a RAND report on regional air operations to discover fuel capacity.[6]

---

[5] When analysis was performed at higher classification levels, more detailed military information was often available with additional fuel storage information for at least some locations.

[6] Hagen , Mills, and Worman, 2013.

Once the total fuel storage of a facility was determined, we calculated the number of fuel storage aimpoints by assuming an average tank size at each location and calculating the number of those tanks necessary to hold the total capacity. The standard tank size used in this study was 25 feet in diameter and held 750,000 gallons.[7] For each tank, we calculate the probability of kill using the standard probability of damage equation:

$$1 - 0.5^\wedge\left(\frac{(tank\ radius + lethal\ radius)^2}{CEP^2}\right), \tag{3.1}$$

where the lethal radius is the distance at which the warhead used can cause damage to a tank and CEP is the circular error probable miss distance of the incoming weapon. A daily fuel resupply rate can be input for each location to capture the varying effects of pipeline, road/rail, or sea-based resupply. One can also imagine actions that could be taken by Red forces that might interrupt the flow of fuel to a base. Strikes against sources of resupply, such as refineries, pipeline, and ports, could be quite important to examine. Furthermore, a successful strike against the pumping infrastructure of the airbase could potentially degrade the base's ability to replenish its fuel stores. For locations that are supplied by sea, strikes on ports or interdiction of vessels while underway are possibilities as well. Outside of military action, adversary governments could also attempt to use political pressure to keep a host country from allowing its refineries to supply American bases with fuel. Further work examining logistics and supply chains would likely offer additional insight into these issues.

Fuel Offload Rates

The AAFIF generally did not provide adequate information about fuel offload flow rates. Not only were the figures inconsistent and seemingly inaccurate, they were also often missing for the airfields we examined. Therefore, we decided to base our numbers on figures provided by U.S. Department of Defense (DoD) maintenance Unified Facilities Criteria on petroleum systems[8] and a Naval Postgraduate School thesis written by Michael Chankij,[9] which dealt with jet propellant distribution resilience on Guam. We also used information from pertinent fueling system manufacturers.

DoD's Standard Type III Pressurized Fuel Hydrant system, found at Andersen Air Force Base, has the potential to move 2,400 gallons per minute (gpm).[10] In this system, pressurized jet fuel stored in operational tanks flows through conduits embedded under parking aprons and is pumped through control valves by hydrant trucks into the aircraft. This system is known as a

---

[7] This could cause errors in cases where bases have a large number of very small tanks or small number of much larger tanks. In the latest update of the model, an average tank size was defined for each location, avoiding this limitation.

[8] DoD, 2003.

[9] Chankij, 2012.

[10] DoD, 2003.

"hydrant loop," with each loop having several fueling points and its own pump house. Andersen Air Force Base, a large permanent base on Guam capable of hosting large forces, has four hydrant loops.

The hydrant loop system represents the final links in a fueling chain that may also be limited by the pipes, tanks, pumps, manifolds, and hydrant trucks that play a role in lifting fuel into an aircraft. The model presented in the Chankij thesis, which was developed by the Air Force Institute of Technology, posits that Andersen's fueling systems fail to meet fuel demand once rates climb higher than 4,500 gpm for the entire base. Since Andersen has four hydrant loop systems, we can assume that each hydrant system unit can supply a maximum of 1,125 gpm. We limited our prime flow rate to this steady-state figure. We also assume that each hydrant loop, or 1,125 gpm of capacity, has nine aimpoints that Red could attack and that would linearly degrade offload rate if destroyed.[11] These fuel offload aimpoints are treated similarly to fuel tanks for calculating damage, except that we assume they are point targets; thus, the equation governing probability of kill is given by:

$$1 - 0.5^{\left(\frac{lethal\,radius^2}{CEP^2}\right)}, \qquad (3.2)$$

where the lethal radius is the distance at which the warhead can cause damage to a node and CEP is the circular error probable of the incoming weapon.

We use this 1,125 gpm flow rate per loop as a reasonable baseline for the hydrant systems found on permanent U.S. Air Force bases and large commercial airports (which may actually have a faster flow rate in some cases), as well as defense installations in Japan, South Korea, and Singapore. Smaller, more remote airfields, such as those found in the Philippines, may not have this level of infrastructure. In these smaller airfields, we assume that fuel trucks, also known as refuelers, would haul JP-8 from storage depots and fill waiting aircraft.

To derive an estimate for lower flow rates, we looked at a series of commercial refueler trucks from a variety of international manufacturers. We set a lower bound on a candidate refueler capacity at 2,200 gallons—enough to completely refuel a typical fighter aircraft in one session. We then took the average flow rate of a sample of the largest non-tractor-trailer refueler trucks (5,000–6,000 gallons capacity) as the upper bound. We assumed that the more remote airfields would not have access to premium, full-size refueling equipment.

Flow rates on modern aircraft refueler trucks in the smaller category range from a low of 200 gpm to a high of 1,004 gpm (the upper range representing a large, twin-hose system from manufacturers in the United States).[12] The average flow rate from our truck sample was 550

---

[11] So, for example, killing four of the nine aimpoints would result in only a 500 gpm offload rate for that loop.

[12] These figures represent the average flow rates found in technical brochures from various manufacturers. Semi-trailer trucks can carry up to 20,000 gallons.

gpm.[13] We sourced these numbers from data provided by manufactures, truck distributors, and service providers such as Garsite (USA), Rampmaster (USA), and Liquip (AUS).

We used AAFIF data and imagery analysis to determine whether the airfields in our set used hydrant systems or refueler trucks. We marked airfields as "equipped" if we determined that they had hydrant systems, we marked them as "semi-equipped" if they had access to lower-flow rate hydrant systems or high-volume refueler trucks, and we marked remote or underserviced airfields as "deficient" if they had access to only smaller trucks. We assigned each base a flow rate of 1,125 gpm, 550 gpm, and 200 gpm per hydrant, large truck, or medium truck, respectively. Many of the airbases had more than one truck or hydrant system that needed to be taken into account. We gave airfields with no refueling infrastructure a zero flow rate.

A significant portion of the airfields relevant to our model have commercial infrastructure. This warranted an investigation into flow rate figures for large commercial airports. According to a study by the International Air Transportation Association (IATA), Heathrow has three hydrant systems[14]—one that serves terminals 1, 2, and 3 (also known as the Central Terminal Area or CTA Hydrant); one that serves Terminal 4 and the Cargo Area; and one that delivers fuel to the Terminal 5 stands. The CTA system probably moves the most fuel, but we can take the average of all systems to determine the flow rate per hydrant system per minute.

The average throughput for Heathrow Airport was over 5,300,000 U.S. gallons per day, or approximately 3,705 gpm. If we divide that total number by the number of hydrant systems, we get 1,235 gpm—a number that is quite close to our Andersen Air Force Base max flow rate per hydrant of 1,125 gpm.

Runways

AAFIF data were our primary source in determining the size and characteristics of runways. After confirming the correctness of these data on a sampling of runways using Google Earth, we input the data directly into the model. AAFIF also provided a load classification number, which indicates the maximum weight that the runway's surface was capable of supporting without damage. This measurement could be used to determine an airbase's fitness for tanker or other large aircraft operations, but it is not currently used by any model calculations.

Shelters

For many airbases, AAFIF's shelter count differed greatly from the number of shelters that could be seen on Google Earth imagery. The reason for this was probably definitional—the set of buildings that AAFIF designates "shelters" is likely different from the set that Red decisionmakers would consider targeting in a conflict. Therefore, we counted the number of

---

[13] If the smaller airfields use fuelling trucks, we would also have to include the time needed to fill the trucks and estimate the number of trucks operating at any given point.

[14] IATA, 2008.

aircraft shelters that were large enough to protect a U.S. Air Force fighter and assigned a single aimpoint to each. This system does not take into account the range of shelter sizes and levels of hardness but, for our high-level model, the addition of such data would not appreciably affect the overall results. We also include the provision for deployable shelters to be added to each location.

ORAM currently uses a single size and hardness of shelter that requires a direct hit with a unitary warhead (ballistic or cruise).[15] We use a formula similar to that used with fuel tanks to calculate probability of kill in these circumstances:

$$1 - 0.5^{\left(\frac{size^2}{CEP^2}\right)}, \qquad (3.3)$$

where size is typically taken as the smallest length or width dimension of the shelter and CEP is the circular error probable for the incoming weapon. Note that here the lethal radius input is omitted because of the demand for a direct hit on the structure. In cases where warheads are large enough or shelters are fragile enough that misses could also result in damage to a shelter, equation 3.1 would be the more appropriate to use. The model currently has inputs for numbers of small, fighter-sized or large aircraft (i.e., bombers, tankers, and ISR aircraft) shelters at each location. In both cases, destroying one shelter equates to destroying one aircraft.

Parking Aimpoints

Inputs on the size and layout of parking areas are necessary to both capture the capacity of each location and accurately calculate the effects of attacking aircraft parked in the open in these areas. The primary measure used by the model to calculate the losses from these attacks is the number of missile aimpoints needed to "cover" the parking areas at each base when using submunition-equipped missile warheads. These warheads dispense a large number of small explosives into the air just prior to impact, resulting in an impact pattern on the ground. The density and lethal radius of the submunitions gives a probability of kill for any aircraft parked within the dispersal footprint. Based on previous open-source RAND work, we are using a theater ballistic missile (TBM) warhead dispersal diameter of 1,150 feet with 825 one-pound submunitions, giving a probability of kill of 0.8 against small aircraft and 0.9 against large aircraft inside the footprint.[16] Cruise missile payloads tend to be quite a bit smaller, with perhaps 200 submunitions, giving a 400 feet x 800 feet footprint size. When working at higher classification levels, additional information is available for real-world missile systems.

To calculate the number of TBM aimpoints for our set of airbases, we initially attempted to use high-level estimations, such as taking the total parking area provided by AAFIF and simply dividing it by the area of the weapon footprint. However, this approach yielded highly inaccurate

---

[15] Red is assumed to use the most effective type of warhead (unitary vs. runway penetrator vs. cluster) for each type of target.

[16] Stillion and Orletsky, 1999. Based on recent work examining the energy required to damage aircraft components, we require a direct hit by a submunition on the aircraft for a "kill."

results upon inspection of real-world locations. Satellite images of the airbases made the reason for this inaccuracy intuitively clear: Depending on the specific geometries of the parking areas, the method nearly always either overestimated or underestimated the number of missiles needed to meet the saturation point. We next attempted to use these AAFIF parking area data in a somewhat more aggregate fashion by calculating a curve fit to the parking area size versus TBM aimpoint data. Figure 3.1 shows an example of such a curve fit for a sample of 10 civil, military, and dual-use bases in the Asia-Pacific region.

Figure 3.1. Fitting TBM Aimpoint and Parking Area Data

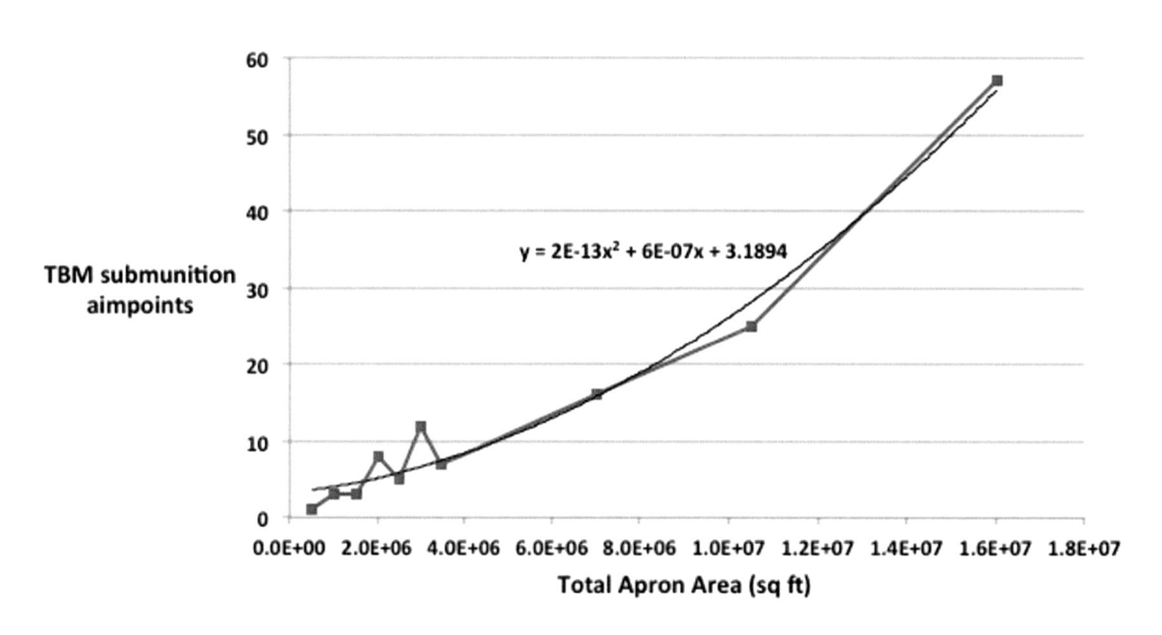

However, we can see that this fit is increasingly poor as base size shrinks, in cases causing errors of factors of two. We also tried to apply other RAND models, but we deemed them unsuitable for our set of airbases because of difficult integration and large impact on run time.

As a result, we ended up simply capturing satellite images of each location, drawing scale footprints for each type of weapon system, plotting these footprints as representational aimpoints, and counting them manually. Although seemingly laborious, we concluded that this method of determining the number of aimpoints was more efficient and accurate than automation, given the limited set of airbases and the variety of parking layouts. For this study, we planned to examine fewer than 100 bases, which allowed us to create an accurate and time-efficient workflow of image capture, scaling, plotting, and counting.

As an example of the targeting process, we examine Reagan National Airport in Washington, D.C., shown in Figure 3.2 with the parking areas highlighted in yellow. The total parking area there, according to the AAFIF, is 3,250,000 square feet. For this example, we will assign a notional 1,600-foot dispersal pattern diameter to an attacking missile. If we simply divided the total area of the parking apron at the airport by the area of the weapon footprint, we would determine that just two TBMs should be sufficient to blanket the parking area with submunitions.[17] A quick glance at Figure 3.2 reveals that that estimation is obviously incorrect, given the narrow, "L-shaped" parking arrangements around the terminals.

Figure 3.2. Reagan National Airport Imagery

SOURCE: Google Earth.

To improve on the that estimate, as seen in Figure 3.3, we first scale the notional 1,600-foot missile cluster munitions footprint to the image using open-source mapping software. Each of the red lines on the image represents the footprint diameter.

[17] The area of the footprint is 2,010,619.3 sq. ft. Thus 3,250,000 sq. ft./2,010,619.3 sq. ft. = 1.62, which we would round up to 2 TBMs required.

Figure 3.3. Reagan National Airport Imagery with Weapon Footprint Diameters Overlaid

SOURCE: Google Earth.

Next, we locate each parking ramp and overlay circular warhead footprints over each, or over multiple ramps if possible. These can also be adjusted to provide some overlap because of CEP and to cover additional structures or infrastructure as possible. Figure 3.4 shows the image with the final aimpoint tally—five TBM aimpoints to saturate the parking areas of Reagan National with submunitions. This simple example also illustrates how submunitions can make TBMs very efficient killers of parked aircraft—every one of the 28 parked airliners at the airport will almost certainly be hit by at least one pound of explosives.

Figure 3.4. Reagan National Airport Imagery with Weapon Footprints Overlaid

SOURCE: Google Earth.

Basing and Beddown Review

Since the aircraft, basing, and deployment information can get fairly voluminous, a review sheet is provided in the model that gives cumulative totals of aircraft by type and location and day-by-day. This sheet is also used to do some simple error checking for the user, including parking area capacity restrictions for small and large aircraft sizes and fuel storage and offload capacities for the current time step. Note that parking restrictions do not limit the model in any way; they are only flagged for the analyst. Fuel restrictions, however, do limit sortie rates, as discussed in the subsections below on refueling and in Chapter Four.

Airbase Repair and Resilience

With these descriptions of airbase capabilities, we can now specify some characteristics that govern their susceptibility and resilience to attack. There are seven basic systems that serve as targets at each airbase: runways, parking areas, fuel storage, fuel offload (hydrants or trucks), shelters, air defenses, and "other." Air defenses will be treated separately later, because they do not affect sortie generation directly, and attacks on parking areas and shelters are treated as a loss of aircraft, as discussed above, rather than a sortie rate degrade.

Runways are one of the most obvious pieces of airbase infrastructure that are required for sortie generation, and attacks against them have received a significant amount of attention from the Cold War era to today. We model up to four runways at each location and calculate the number of "cuts" in all of these runways required to prevent sorties. A runway cut is simply sufficient damage (typically craters in the surface) across the width of the runway to prevent an aircraft from crossing that point during a takeoff or landing. An example of a cut is shown in Figure 3.5. Smaller aircraft, which require shorter distances to takeoff and land (known as a minimum operating surface, or MOS), require more cuts than large aircraft to prevent them from taking off or landing. Similarly, long runways must be cut in multiple points to prevent aircraft from operating between the damaged areas. Airbases with long parallel taxiways or even large parking areas may also require these to be cut to prevent flight operations.

Figure 3.5. Illustration of a Runway Cut

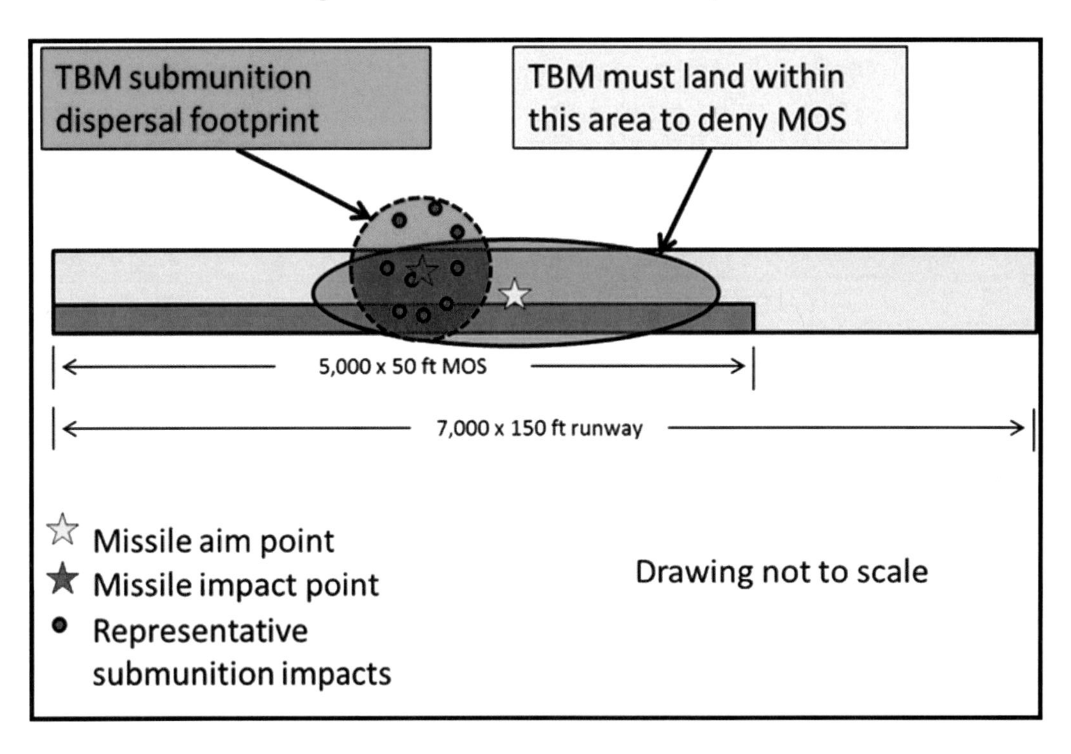

Once the total number of cuts needed to fully close each location is known for small and large aircraft (the MOS input is currently set to 5,000 feet for small and 7,500 feet for large aircraft), we can calculate the results of an attack using the basic probability of kill = 1 – (1 – SSP_K) ^ N, where SSP_K is the single shot probability of kill and N is the number of impacting weapons per required cut point. To calculate the SSP_K, we approximate the intersection of the missile's submunition footprint and the runway as an ellipse and use the formulation for probability of hitting an elliptical target:

$$\sqrt{\left[1 - e^{\left(\frac{-SMA^2}{2 \times sigX^2}\right)}\right] \times \left[1 - e^{\left(\frac{-SMI^2}{2 \times sigY^2}\right)}\right]}, \tag{3.4}$$

where SMA is the semi-major axis of the ellipse, sigX is the mean standard deviation of miss distance along the major axis, SMI is the semi-minor axis, and sigY is the mean standard deviation of miss distance along the minor axis.[18] If we assume a circular miss distance and know the CEP, then

$$sigX = sigY = \frac{CEP}{1.774}. \quad (3.5)$$

The repair time for each runway cut is a user input, usually set to between eight and 24 hours.[19] The lower end of this broad range assumes a capable repair team with state-of-the-art materials; more basic locations may require much longer repair times.[20] Note that we include any time for damage survey, assessment, and explosive ordinance disposal in this category of "repair time." These elements of repair time are typically fixed, so even improved capabilities are likely to have at least a few hours of "repair time" input. The second runway repair input is the total number of runway cut repairs that can be accomplished. Since most locations have a limited supply of repair materials, such as gravel, concrete, and matting, it is important to capture the effects of depleting these materials through repeated attacks on runways. Obviously, if the user wishes to assume resupply is available, this input could be set quite high. Once this number of repairs has been exceeded, the base is assumed inoperable for the remainder of the campaign and will fly no further sorties.

Attacks against fuel storage are assessed to examine whether on-base fuel storage can meet the demands of the current time step. If it cannot, the sortie rate is reduced accordingly. We first calculate the fuel needed to fill each planned sortie to capacity and compare to the current fuel storage status. Fuel storage is degraded through damage to storage tanks and is computed as a linear degrade—if half the tanks are damaged, there is half the baseline fuel storage available. Daily fuel resupply, which is an input for each location, is then added to the remaining storage to determine fuel available at the current time step. If the available fuel storage is sufficient to fly 24 hours of sorties, there is no degradation. If, however, the storage falls below that level, each time step is degraded by the fraction available. So, for instance, if there is only 75 percent of the day's needed fuel available, each time step in the next 24 hours will fly 75 percent of the planned sorties. This prevents no sorties from flying at time steps later in each 24-hour period, because the earlier ones consume all the available fuel. There is also an input for fuel storage repair, currently set to 10 percent per day. This may be quite optimistic for some types of attacks. For example, if 20 percent of the fuel storage were destroyed on Day 1, it would be back to 100 percent at the beginning of Day 3. It would not be difficult to make this repair rate a function of damage level, however, obtaining more realistic information on fuel storage repair rates, especially at various types of bases, would be an important prerequisite to adding this level of detail.

---

[18] For the derivation of this equation, see Przemieniecki, 2000, pp. 45–46.

[19] Runway repair times for unprepared locations or perhaps for aircraft carriers could easily be as long as 48 hours, however. There appears to be little analytic data available on these more "unusual" locations.

[20] See, for example, Mellerski and Rutland, 2009.

Damage to fuel offload capability and the resulting sortie degrades are treated in a fairly similar fashion. Loss of fuel handling aimpoints are converted to loss of offload rate (in terms of gallons per minute) using the nine aimpoints per 1,125 gpm of capacity factor discussed above. At each time step, the needed offload rate is determined by dividing the fuel needed at each base by the length of the time step. If the offload rate cannot meet this demand, sorties at the affected location are scaled accordingly. Note that we do not want to double-count sortie penalties in the case of limited fuel storage and fuel offload, so the minimum allowable sortie capacity of the two factors is used. Fuel offload has a repair rate per day as well, currently set to 25 percent to reflect a presumed greater ease of repair (and possibly greater scope for workarounds) compared with fuel storage tanks. However, this may not always be the case—for example, if fuel tanks are easily patched but fuel hydrant systems are hardened or underground and rely on valves and pumps that are not easily replaced. As with fuel storage, more research into this area is needed.

The final category of airbase target, "other," is simply a placeholder for airbase infrastructure types that an analyst may consider critical and want to include. For example, we earlier highlighted the likely importance of maintenance, supply, and personnel in supporting sortie generation; this category could be used to capture some of those effects, if inputs such as number of aimpoints, probability of kill versus various weapon systems, and the relationship between the loss of aimpoints and sortie generation were known. Currently, this category is set to have 25 aimpoints at each base, weapons have a 15-meter lethal radius against them, and sortie rate is scaled linearly with the number of "other" targets lost.

Issues that likely deserve further attention across these types of aircraft carrier and airbase infrastructure categories are the repair capabilities and capacities. How long would holes in a carrier flight deck take to repair? What are real-world repair times for aviation fuel storage tanks and offload hydrant systems? What about for fuel bladders? Could parking areas be easily repaired with flexible mats? Are carrier catapult and landing systems uniquely vulnerable to damage? Would attacking personnel in barracks or tent cities have a catastrophic effect on all other repair types? These questions and many more have received little attention by the military services and the analytic community since the Cold War, and yet may have a very large effect on accurately measuring airbase resilience.

Refueling Orbits

One of the key factors included in ORAM is an accounting of aerial refueling processes, including tanker sortie rates, refueling track locations, and offload versus range of various tanker types. The objective of this module is to measure whether aircraft sorties from each location have sufficient fuel to reach their destination and return, and, if not, to limit sorties from those locations to executable levels.

Three types of refueling aircraft are provided to choose from—KC-135R, KC-10, and KC-46—with fuel offload calculated as a function of range for each type. However, because a single type must be used throughout the theater, a representative tanker can be defined that reflects the capabilities of a mixed fleet.[21] Once a tanker type is chosen, the next step is to locate the various refueling orbits (up to 20 are allowed). During model development, we explored several options to automate this process, but in the end we determined that simply letting the user enter the latitude and longitude of logical orbit locations was the best trade-off between transparency, model run speed, and analyst workload. Several tools are provided to help with this process. First, as mentioned above, the number of needed refueling orbits is calculated for both small and large aircraft for each basing location. This first-order calculation assumes that aircraft will need refueling orbits spaced at 50 percent of their maximum range (i.e., they never get below 50 percent onboard fuel). The second tool is a table giving the distances from every basing location to every refueling orbit. This allows the user to visually check whether the entered orbit locations are capable of supporting every location.

The next required input is the matching between refueling orbits and basing locations, since aircraft flying from each location will not be using every orbit. This entry table (sample shown in Table 3.1) is also provided with user assistance by highlighting locations (green in Table 3.1) with small or large aircraft and the number of orbits required to support each type (using the same 50 percent rule of thumb mentioned above). For example, the user may be shown that Base X requires three orbits for small aircraft and none for large aircraft (either because none are based there or because it is close enough that no refueling is needed) and then can choose which three orbits best support that base. Orbits can be "attached" to as many locations as desired. Although the model highlights for the user how many orbits are recommended, there is no error checking to prevent aircraft from flying unrealistically long ranges if orbits are spaced too distant. The fuel and tanker demands would be calculated correctly for this case, and the sortie would be allowed to fly its mission.

[21] We have currently limited the model to a single type of refueling tanker, because otherwise additional inputs would be needed to define which tanker types would be associated with each refueling orbit. If this degree of specificity were desired, it would not be difficult to add. Offload versus range as currently specified in the model for the KC-10 and KC-135 are taken from Air Force Pamphlet 10-1403, 2011, and preliminary values for the KC-46 from personal communications with Major Glenn Rineheart, Air Force Operational Test and Evaluation Center.

Table 3.1. Sample of Location-Refueling Orbit Matching

Name	Small / Large needed	Orbit 1	Orbit 2	Orbit 3
CSG 1	1 / 0	1		
CSG 2	1 / 0	1		
CSG 3	1 / 0	1		
CSG 4	0 / 0			
CSG 5	0 / 0			
Andersen	4 / 1	1	1	1
Atsugi	0 / 0			
Darwin	0 / 0			
Fukuoka	0 / 0			
Futenma	1 / 0	1		

Once the orbits have been located and matched with each location, the demand and supply and any required limits can be calculated for the current time step. We first calculate fuel storage and offload capacity needed at each base to fill each sortie with fuel; this calculation is used to ensure that enough fuel handling is available to launch the time step's sorties, as discussed above. Next, the total airborne demands of each location are calculated, using the distance, fuel flow, and any loiter time for each type of aircraft.[22] The time spent loitering is currently set to consume 75 percent of cruise fuel flow, and an overall 15 percent safety margin is applied as well. Once the total demand for each location is known, this is divided evenly among the refueling orbits supporting that location. This is somewhat of an approximation, since the actual spacing of orbits may result in higher and lower relative demand; however, this effect should not be large, given the level of aggregation in the model. These location-specific demands are then summed to give the total fuel demand needed at each orbit.

To calculate the ability of the deployed tanker force to meet that demand, we calculate the number of sorties available from each tanker basing location when flying to each orbit. These sorties are weighted based on the relative demand of each orbit. So, for instance, if Orbit 1 requires 50 percent of all the airborne fuel needed for this time step, it will receive 50 percent of all the tanker sorties flown. This weighting allows the orbits to be somewhat "self-equalizing," since, for example, if demand drops at a particular orbit because of basing attacks, that orbit will in turn receive fewer tanker sorties. Note that every location with tankers supports every orbit location. This is also a simplification, given that typically one or two bases would likely provide dedicated support to a subset of the refueling tracks. However, this assumption is necessary to provide robustness to the refueling system. If one or two locations that support an orbit are shut down because of attack or lack of storage, then that refueling orbit would not be supplied and potentially the entire air campaign could shut down. By allowing all tanker locations to supply all orbits, this likelihood is reduced.

---

[22] We do not include tankers in this calculation. This implies that tankers are not allowed to refuel other tankers, which may be an interesting capability to explore in further work.

The final step is then to compare the supply provided to each refueling orbit (the number of tanker sorties times the offload available at the required distance) to the demand. This is calculated as a fraction of the demand, and, for each location, the minimum fraction of each of the needed orbits is used as the sortie-scaling factor. So, for example, if Base X utilizes Orbit 1 and Orbit 3, and Orbit 1 can only be supplied with 75 percent of the needed fuel and Orbit 3 is supplied with 100 percent of the demand, Base X will be limited to 75 percent of the sorties it might otherwise have flown for this time step because of the limitation at Orbit 1.

Note that we have to be careful about circularity problems here, since demand for fuel is a function of the number of sorties that are being flown, which, in turn, is a function of the other systems on the airbase as they come under attack. To avoid this problem, we first calculate the results of all the other attacks on an airbase, such as runways, fuel storage, and parking, and use that result as the tanker demand. It is this potentially degraded sortie demand that is used to calculate tanker demand, which is then degraded further, as discussed above, if necessary.

Missile Defenses

Missile defenses of four different types can be deployed to each location, with two layers for ballistic missile defense and two for cruise missile defenses. For example, these might include Aegis SM-3 and Patriot as outer and inner TBM layers and Patriot and short-range air defense (SHORAD) systems for cruise missiles. There is not a day-by-day TPFD for defenses; they are generally assumed to be in-place prior to the start of the conflict. However, if Excel formulas are used instead of a simple input value, then simple rules such as "if day < 3 then num_defenses = 0, else num_defenses = 2" can create a flow of assets into the theater.

The capability of each type of missile defense is defined by a set of inputs, including:

- detection range against a 1 m^2 radar cross section target (against cruise missiles only)
- maximum intercept range (cruise missiles only)
- minimum intercept range (cruise missiles only)
- interceptor speed (cruise missiles only)
- command and control delay between salvos
- interceptors to fire at each incoming missile
- probability of decoy discrimination
- probability of kill of each interceptor (single shot probability of kill [SSP_K])
- maximum number of simultaneous engagements
- number of ready interceptors at each site
- whether decoys need to be distinguished during a second salvo (against TBMs only)
- effectiveness degrade versus intermediate-range ballistic missiles (IRBMs) (TBMs only)
- effectiveness degrade versus medium-range ballistic missiles (MRBMs) (TBMs only).

When Red fires a salvo of TBMs, their numbers are first reduced by a reliability input. The remaining missiles, because they are targeted at each location, are then engaged by the defenses protecting that specific location. The model "fractionates" the defenses that are specified as covering multiple targets. So, for example, if a defense is protecting two locations and they are both attacked, they would each receive a fraction of the defense weighted by the number of missiles fired at each. If, however, only one of the two locations is attacked, it would receive the full capability of the defense. This fractionation simply scales relevant inputs, such as the number of ready missiles and the number of simultaneous engagements. Similarly, when multiple sites of the same type are at a single location, those variables are multiplied as well. The TBM defenses are divided into outer and inner layers, so the outer-layer defenses engage first. These outer-layer defenses either engage with a single salvo (the number of interceptors to be used against each incoming missile is user specified) or as two "shoot-look-shoot" salvos. This capability can be turned on or off by a flag, which is typically used to define the presence of an early warning radar forward of the air defense itself.[23]

Included in the list of targets to be engaged by the outer layers are missile decoys. Their number per incoming missile is input as a characteristic of the Red weapon system, and Blue's ability to discriminate them from real warheads is an input for each defensive system type. This input is defined as the probability of decoy discrimination. For example, if a salvo of ten missiles deploys four decoys each, there would be 50 potential targets seen by the air defense, only 20 percent of which are real warheads. However, if the decoy discrimination input is set to 75 percent (very good performance), then the defense would actually engage only 25 percent of the decoys (ten out of 40), along with the ten real warheads; thus, 50 percent of the engaged targets would be real.[24] The calculations in the model are somewhat more complicated, since there are three types of ballistic missiles, short range (SRBM), medium range (MRBM), and intermediate range (IRBM), all of which may have different numbers of decoys per warhead, but the principle remains the same.[25] Once the numbers of decoys and warheads to be engaged are known, the maximum number of simultaneous engagements input is applied, and the user-specified number of interceptors per target is launched at the decoys and warheads in the correct proportion. The model also includes an input for each defense type that scales the number of engagements of each type versus each incoming missile type. This is typically used to give some types of defenses less capability against longer-ranged, and hence faster, ballistic missiles.

Actual kills of decoys and warheads are assessed using the standard

$$1 - (1 - SSPK)^{N} \tag{3.6}$$

[23] Defenses often need additional radars forward of their position to maximize the engagement envelope against ballistic missiles, because their own indigenous radar can be limited by constraints such as the curvature of the earth, radar power, and search volume.

[24] This is somewhat of a simplification of the real world, where two types of errors are actually possible—decoys can be mistaken for warheads (the error modeled here), but warheads can also be mistaken for decoys. The model currently ignores this second type of error, and thus assumes defenses are more capable than they might actually be.

[25] There is currently no capability for distinguishing different types of decoys at different rates, if, for example, SRBM decoys were easier to distinguish than IRBM decoys.

formulation, where here the SSP_K is a user-specified value and N is the number of interceptors per target. Note we do not apply any missile defense failures because of guidance or interceptor fly-out problems, so the analyst must use the SSP_K input to account for any real-world issues here, as well as other countermeasures, such as electronic attack or missile maneuvers. Surviving decoys and warheads are then passed to the second salvo if shoot-look-shoot is enabled; otherwise, they pass to any inner-layer defenses. There is also an input to enable a capability to ignore decoys on the second salvo to capture the effect of decoys, such as balloons, that are effective outside the atmosphere but not usable at lower altitudes.

The second-salvo or inner-layer calculations are adjudicated in the same way, with the maximum number of simultaneous engagements and number of interceptors per target employed against decoys and warheads as appropriate. The only difference is that shoot-look-shoot engagements for the inner-layer defenses can be enabled by the presence of outer-layer defenses (they would acting as a cue in this case) in addition to the forward-based radar discussed above. Thus, in total, up to four independent engagements can be employed against incoming ballistic missiles, or as few as one if only a single defensive layer is available and shoot-look-shoot is not available. Any surviving warheads after this engagement process are then used to calculate damage to the various airbase targets or to the defenses themselves.

Defense against air- and ground-launched cruise missiles is similar, with outer and inner layers of SAMs (such as a long-range Patriot and a medium-range Hawk or a man-portable infrared missile) and multiple salvos; however, the rules governing their use are somewhat different. For cruise missiles, we specify the altitude and radar cross-section of the incoming cruise missiles and the altitude and radar performance of a cruise missile defense sensor and calculate the initial engagement range. Using the time required for each engagement and the user-specified speed of the cruise missiles, along with command-and-control delays, we allow multiple salvos for the outer cruise missile defense layer until a minimum range is reached, up to four in total. The inner layers of cruise missile defense are modeled as a short-range SAM with multiple salvos as described above. At the outermost level, we allow the user to specify a number of aircraft at each location to maintain orbits and perform the cruise missile defense mission. There is no modeling of the dynamics of these engagements; the user simply specifies the number of cruise missiles each orbit will shoot down. The user can specify how many two-ship defensive orbits are protecting each location and the model will calculate how many sorties will be "lost" to this mission. These incoming missiles are removed before encountering any of the ground-based air defenses. A further implication of this airborne cruise missile defense is that these sorties do not contribute to any of the operational measures of effectiveness (MOEs). Given that it can require 6–12 aircraft to keep a two-ship of fighters permanently on-station, the analyst must keep the trade-offs in mind between providing a heavy cruise missile defense and having more aircraft sorties available for other, presumably offensive, missions.

For both the TBM and cruise missile defenses, a final layer is provided that is simply modeled as an independent sensor, a number of "bursts," and a probability of kill per burst. This can be used to account for the presence of traditional anti-aircraft guns, but also to capture new defensive technologies, such as rail guns or directed energy systems, at a low-fidelity level.

Both layers of TBM defense and the outer layer of cruise missile defenses are also vulnerable to missile attack themselves.[26] The Red missile plan (discussed in detail in the next chapter) can select each defense type as a target, and each type has a set of inputs determining its vulnerability. A site has a set of critical nodes (currently set to four for all types), a desired probability of kill (now 0.9), and a lethal radius for weapons against it (currently 100 meters). For example, with these inputs and a weapon with a 50 meter CEP, it would require just over four impacting missiles to destroy one Blue air defense site. Partial kills result in partial loss of capability, because the site is fractionated and interceptors are proportionally lost. This may somewhat overstate the survivability of an air defense site given the single point of failure in the target tracking radar.

---

[26] The system forming the inner layers of cruise missile defense was not considered likely to be attacked due to its mobility and low value. Aircraft used as cruise missile defenses are obviously vulnerable as well—they are affected by basing attacks, just as other aircraft are.

4. Modeling Red Capabilities

As mentioned in Chapter Two, Red has somewhat fewer input requirements than Blue, mainly because of the lack of infrastructure elements. Most of the Red inputs revolve around missile attributes and the plan for determining how many are fired, at which locations, and when.

Red Knowledge

One of the key elements for guiding a Red attack strategy is access to intelligence about Blue's ongoing operations. Where is Blue operating from? How damaged is a particular base? Is it still defended? Have additional aircraft arrived at a location? Are Blue sorties limited by fuel, runways, or aerial refueling? These are all questions a Red commander would be asking, and they are questions that are unlikely to be answered with the "ground truth." More likely, only delayed, incomplete, and partially incorrect information would be available.

To capture some of these real-world effects in the simulated Red planning process, the model operates using information that has user-specified delays and errors applied to it, if desired.[1] The analyst can specify how old, in terms of time steps, the various data elements are. So, for instance, the number of sorties Blue is flying may be taken from several hours or days in the past. For periods earlier in the campaign than the delay, preconflict information is used instead.

There are currently 11 elements of data for each location captured as part of this intelligence process, although this could be easily expanded:

- number of outer and inner TBM and outer cruise missile air defense sites
- number of aircraft shelters
- number of tanker sorties flown
- number of fifth-generation fighter sorties flown
- number of fourth-generation fighter sorties flown
- number of long-range strike sorties flown
- runway repair capability remaining
- current MOS length
- fraction of fuel storage operational.

These elements are first calculated as ground truth for the appropriate time period, then errors are applied. These errors can be of any type allowed by Excel formulas. As some examples, we have experimented with setting some of the values to zero for certain locations, adding or subtracting percentage values as errors, fixing the data to starting values, etc. There are few

[1] Note that there is no parallel module for Blue operations. Since Blue does not have the same explicit targeting challenge that Red faces, we have not implemented any type of information gathering and command-and-control logic for Blue. However, one possible example of a need for this might be Blue's decisions on where to place arriving forces or how and when to move forces away from damaged locations.

restrictions on the types of manipulations that could be performed. However, there are two caveats. First, setting values to zero may have different meanings in different contexts. For instance, zero TBM defenses implies that the base is undefended (when it may actually have defenses in place) but still fully functional, while setting the MOS to zero implies the runway is closed and so is not flying any sorties. Similarly, random errors must remain invariant from case to case, because no repetitions are being done.[2] There is also no built-in way to have error size or type change from day to day, although, again, some simple Excel if-then-else logic could be easily entered in place of scalar error values.

Missile Capabilities

As mentioned above, there are five types of attacking missiles defined—three types of ballistic missiles (SRBM, MRBM, and IRBM) and two types of cruise missiles (air- and ground-launched). For the ballistic missiles, the user defines the failure rate, accuracy in terms of CEP, the dispersal radius of submunition warheads for parking area and runway attacks, probability of kill within the footprint against aircraft, number of decoys per warhead, the maximum range, and the latitude and longitude of three possible firing locations. For example, two of the three launch locations could be in the adversary's territory and the third at a weapon release point closer to Blue's basing.[3] Note that the warhead type used for each type of target is effectively chosen by the lethal radius inputs. So, for example, if a user inputs that 100 TBMs are needed to cover a parking area, they are likely assuming a unitary warhead, while if 20 is input instead, the user has likely chosen some type of submunition. Similarly for the lethal radius of TBMs and cruise missiles against point targets like shelters or fuel tanks. The model simply makes a calculation for each target type. This approach could be thought of as employing the best warhead against each target type, assuming that the user input the most effective warhead type for each target type. This is likely a realistic assumption for Red, but one shortcoming is that it cannot account for Red having a limited quantity of some warhead types.[4] The range and firing location information is used to calculate whether the missile has sufficient range to reach each basing location. This information is provided to the user but is not used to prevent firings—the analyst can fire any missile type against any location if desired. Cruise missiles have three additional inputs: altitude, radar cross-section, and speed. These are used to determine the maximum

---

[2] It would actually not be difficult to repeatedly run the model; a simple macro that repeatedly calls the current time advance macro and saves the outputs would be the only necessary addition. Run time would obviously lengthen accordingly.

[3] One issue that may be useful to capture in follow-on work is the effect of forward air defenses against the aircraft launching cruise missiles. Killing them, as opposed to each individual cruise missile, is likely to be a much more efficient use of fighter aircraft.

[4] The probability of kill against aircraft within the submunition footprint is a user input, because of wide variance in missile payload sizes and submunition types. One simple way to approach the problem for the analyst is to choose a submunition density (and, hence, footprint size) for small and large aircraft that ensures any aircraft in the footprint will be hit with at least one submunition. Then, the probability of kill can be set to 1.0.

engagement range and the number of subsequent salvos of cruise missile defenses. There are both air- and ground-launched cruise missiles available, as well as an "other munition" type, primarily for use when airbases come under direct attack by aircraft.

Attack Strategies

There are two different inputs for building the firing plan for ballistic and cruise missiles. First, for each missile type, there is a table of location versus time step where the number of missiles to be fired at each step can be entered. This is obviously quite simple and transparent and allows the analyst complete control over Red's strategy. The analyst can use this table, the calculation of locations that are in range, Red's intelligence information (discussed above), and inputs on allowed missile firings per day and total inventory to create an automatic firing plan, if desired. For example, simple Excel formulas have been used to create a hierarchical set of checks, such as the following:

- Is this location in range of this missile type?
- Are sorties flying from this location?
- Are missiles available in the inventory?
- Are missiles available this time step to be fired?
- If all questions are answered yes, fire 25 missiles at this location.

Note that this is a very simple set of logic; additional checks, such as what type of aircraft are operating from the location and whether other types of missiles have already been fired at the location, can all be used quite easily to create a fairly automated Red attack planner. In addition, all the information used to make these decisions does not have to be ground truth; it can be based on the delayed and incorrect intelligence information discussed above.

The second choice required in building Red's attack strategy is the allocation of fired missiles against the various Blue targets. Recall that there are seven different target types to choose from: aircraft shelters, fuel storage, fuel offload, runways, parking areas, air defenses, and "other." The basic allocation inputs are entered as the fraction of missiles in a table of locations and targets. Thus, the analyst can specify that at Base X, 25 percent of any missiles fired at it should go against shelters, 20 percent against defenses, and 55 percent against fuel. There are separate entries for ballistic and cruise missiles because of their differing warheads and accuracies. In addition, there are three different campaign phases available. The user can determine the ending day of each phase, and a different missile allocation can be entered for each. This allows even a simple targeting scheme to have some dynamics, such as attacking defenses first, then attacking runways to trap aircraft, and then attacking parking areas and shelters to destroy those parked aircraft.

As with the number of missiles to fire, these allocation rules can be simple inputs or formulas using various types of information to set the allocation in a more dynamic fashion. For example, the baseline ORAM configuration includes checking whether air defenses or shelters are present at a base before allocating missiles against them. This will allow attacks on those elements to continue until they destroyed (or degraded to some desired level) and then other types can be engaged. The allocation rules also currently include stopping runway and fuel attacks if no sorties are flying and attacking only parking areas at bases hosting specific types of aircraft (for example, fifth-generation fighters or large aircraft, since their size and lack of shelters makes them more vulnerable). The level of complication and amount of Red intelligence used is truly limited only by the expected Red doctrine and the imagination of the analyst.

5. Sample Outputs Using ORAM

Now that the discussion of inputs into the Blue and Red sides of ORAM is complete, we run through the calculations that occur as the model is advanced from time step to time step, discuss the outputs that are generated, and give some sample results. The companion report describes several complete case studies using ORAM with input data at higher classification levels.

How Model Results Are Calculated

When the time step is advanced, the model performs a full series of calculations and generates a complete set of output data. It begins by updating the number of aircraft of each type present at each location, based on any losses from the previous step and any changes from the beddown because of either new deployments to the theater or relocations of aircraft within the theater. Similarly, the number of air defenses, their interceptors, shelters for aircraft, fuel, and Red's missile inventories are updated as well, including increased capacity from the previous time step due to repairs. ORAM then calculates Red's missile attack, using the firing and allocation logic just discussed in Chapter Four.

Adjudicating air defense effectiveness occurs next, with the missiles fired at each location reduced in number as appropriate. Note that there is no preferential targeting by the air defenses based on the target of the missile. For example, missiles heading toward aircraft shelters cannot be engaged instead of those heading toward fuel storage, although this lack of preferential targeting may be fairly realistic, given many targets' close proximity and uncertainties in the point of predicted impact. Instead, the surviving missiles are allocated against the targets in the same proportion in which they were fired. For each target type, a weighted average missile CEP is calculated, using the number of arriving missiles by type. The weighting scheme for each type of target uses the formula for the weighted mean:

$$CEP_{avg} = \sqrt{\frac{Num_1 + Num_2 + \cdots + Num_n}{\frac{Num_1}{CEP_1^2} + \frac{Num_2}{CEP_2^2} + \cdots + \frac{Num_n}{CEP_n^2}}}$$

where the Num_i are the number of each type of missile (SRBM, IRBM, etc.) arriving at specific target type and CEP_i are the individual CEPs of each missile type. For instance, if a total of ten missiles with CEP = 10 meters and five missiles with CEP = 20 meters survived the air defenses and are targeted at aircraft shelters at all locations, the CEP used to determine aircraft shelter kills at all locations would be

$$\sqrt{\frac{10+5}{\frac{10}{10^2}+\frac{5}{20^2}}} = 11.55\ meters$$

The calculation is done for each of the seven target types.

Once the number of impacting missiles is known, their effects on each aircraft carrier and airbase are calculated. The number of aircraft shelters, fuel storage, fuel offload, and "other" target elements destroyed are calculated using the equations discussed in Chapter Three. The fraction of the parking area struck is a simple calculation of the number of arriving missiles divided by the number needed to completely cover the base. Runway closure times are calculated by first determining what fraction of the cuts needed to completely close all the runways were accomplished. Next, this fraction is multiplied by the time to repair a single runway cut.[1] As an example, if a base required two cuts each on two runways, but only enough missiles impacted to produce two cuts, 50 percent of the cuts were accomplished. If the time to repair were input at eight hours, then the base would be considered closed for $0.5 \times 8 = 4$ hours.

Even though there are always operable runways available in this example, the model applies a sortie rate degrade and repair time to account for tasks such as damage assessment, explosive ordnance disposal, and replanning flight operations.[2] Even in the most extreme case of all four cuts being accomplished, the base would be closed for eight hours—the time to open a single cut and create a single MOS. Note that this implies the assumption that bases are not normally operating at full capacity (i.e., they do not need to launch sorties from every runway every few minutes). In general, this is almost always the case, especially at military airfields; thus, this assumption is justified.[3] To ensure that all possible Red attacks are appropriately accounted for, it is important to run with a short enough time step (eight hours is typical) so that Red can re-attack as necessary.

To assess aircraft lost in parking area and shelter attacks, we first determine how many of the aircraft assigned to the base are actually on the ground. The user can specify the fraction of aircraft expected to be airborne at any given time (typically between 40 and 60 percent, depending on the length of the missions being flown). There is also a user input for the number of aircraft able to be "flushed" (or quickly launched from the base) before the attacking missiles arrive. These would typically be a small number of aircraft and pilots kept near the runway in a

---

[1] The model does not currently include degrades to account for repair crews or equipment themselves coming under attack. This could be an important feature to capture in the future.

[2] Fairly ineffective runway attacks could produce closure times that are shorter than the fixed elements of runway repair times, such as survey and explosive ordinance disposal. To avoid this, the user could utilize the "min" function in Excel to ensure that runway closure times were never less than a desired duration.

[3] For instance, if a base with two runways launched a sortie every three minutes from both runways, it would total 960 sorties per day. At typical sortie rates of 1.5 per aircraft per day, this would require 640 aircraft at the base. There are few, if any, bases with sufficient parking for even half this number of aircraft, which totals approximately half the entire U.S. Air Force combat aircraft inventory.

state of readiness that would allow them to get airborne within the warning time of an attack. For locations near the threat, where warning time may only be a few minutes, this may even require the aircraft to have their engines running. These aircraft are kept safe from being killed on the ground from shelter or parking area attacks, but they are removed from the sortie count that contributes to operational MOEs.

With the number of aircraft on the ground known, they are preferentially parked in shelters as available, with the remainder parked in the open. The first set of losses is then the minimum number of shelters destroyed and the number of aircraft in shelters. Since the remaining aircraft are all parked in the open, those losses are calculated by multiplying their number by the fraction of parking area covered by missile footprints and the probability of kill inside each footprint. Note that this assumes aircraft are evenly parked across all available parking areas—i.e., they not concentrated. There is also a user input for any types of additional aircraft losses, such as on missions or on the ground because of adversary special operations forces.

With these results, we can now make the first calculation of sorties to be launched. The basic calculation is the number of aircraft available times the sortie rate for the time step. The number of aircraft available are those deployed minus the losses, those on cruise missile defense, and those that were flushed. The effective sortie rate is calculated by taking the baseline sortie rate for each location and degrading it because of damage to runways and the "other" category. These degrades are the hours the runway is closed for small and large aircraft divided by the time-step length (so, for instance, if the runway is closed for four hours during at eight-hour time step, the sortie rate will be one-half the baseline value) and by the fraction of "other" aimpoints surviving (similarly, if 50 percent of the aimpoints are currently destroyed, there will be a 50 percent sortie rate degrade). These two types of degrades are additive. This effective sortie rate then sets the number of sorties used for the fuel calculations, both in terms of demand and for supply by tankers.

As discussed earlier, if insufficient fuel storage or fuel offload capacity is available, the sortie rate is reduced further to the minimum of these two constraints. These degrades also apply to tankers. This sortie count is then used for the airborne refueling calculations, which may, in turn, reduce the number of sorties further at each location if insufficient supply is available. This final total is the number of sorties available for operational missions.

At each time step, a wide variety of MOEs are reported, and it is not difficult for users to add additional ones they may find necessary. The model currently includes counts of sorties flown by aircraft type and by location for each time step, as well as aircraft lost each step by type and cumulatively by location. There is a running tally of the number of aircraft of each type present in-theater as well. More detailed outputs focused on missile attacks include the number of Red missiles fired by type at each step, as well as Blue air defense sites remaining (which is further broken out by sea and land basing) and the number of interceptors fired. To allow better insight into the reasons behind sortie limits, there is an output table of sortie rate degrades for each location at each time step, as well as a similar table for any airborne tanker degrades.

There are also several operational metrics reported, using the user-defined allocation of sorties by aircraft type discussed in Chapter Three. Recall that the mission types currently defined are number of DCA orbits, strike packages, long-range strike packages, and ISR orbits.

Baseline Case

To illustrate some of the model behavior and output reporting, we show here some results with a simple notional case. We begin with a beddown of approximately 560 aircraft, including fighters, bombers, ISR aircraft, and tankers. Table 5.1 details the totals by types and by 11 locations. Here we are assuming all the aircraft are in place before the conflict starts and that no additional aircraft arrive.

Table 5.1. Aircraft Beddown Used in Example Case

	Total	Type								
		Air-to-air	Multi-role 1	Multi-role 1	Carrier	Bomber 1	Bomber 2	ISR 1	ISR 2	Tanker
CSG 1	40				40					
CSG 2	40				40					
CSG 3	40				40					
Base 1	120	24	24			12	24	24	12	
Base 2	36									36
Base 3	48			48						
Base 4	48	24	24							
Base 5	48		48							
Base 6	36									36
Base 7	36									36
Base 8	72					12	24	24	12	
Total	564	48	96	48	120	24	48	48	24	108

These bases are equipped in our example with several resilience measures as well, including aircraft shelters, runway repair, and air defenses. These elements, and the distance and baseline sortie rates for each, are summarized in Table 5.2 for our 11 locations.

Table 5.2. Basing Location Details

	Distance (nm)	a/c at base	Small a/c sortie rate	Large a/c sortie rate	Fuel storage (gal)	Parking area (ft^2)
CSG 1	300	36	1.6	0.0	350,000	250,000
CSG 2	300	36	1.6	0.0	350,000	250,000
CSG 3	300	36	1.6	0.0	350,000	250,000
Base 1	1,500	120	1.1	1.2	81,000,000	5,000,000
Base 2	1,500	36	1.1	1.2	1,000,000	6,000,000
Base 3	1,000	48	1.5	1.3	5,000,000	5,000,000
Base 4	300	48	2.0	1.3	10,000,000	6,000,000
Base 5	1,500	48	1.1	1.2	10,000,000	4,000,000
Base 6	1,000	36	1.3	1.3	5,000,000	6,000,000
Base 7	2,000	36	0.0	1.1	1,000,000	15,000,000
Base 8	3,500	72	0.0	0.5	2,000,000	4,000,000

The "distance" input is calculated using the location of the base and the destination of the sorties from that base (which can be specified individually for each location) to allow calculation of such variables as flight duration and fuel consumption. The "a/c at base" column gives the total number of aircraft at each location; the model actually tracks each aircraft by type as well. Using this, sortie rates for small and large aircraft are computed based on typical U.S. Air Force planning factors for each base, although aircraft of both types may not be present at every location. Fuel storage and parking area inputs are taken from the AAFIF database or other sources as available.

Table 5.3 also provides detail on the resilience measures available at each location. The "shelters" input is a count of small aircraft shelters. No large aircraft shelters are available in this example. These are used for aircraft parking according to the user's prioritized list of which aircraft types should be sheltered first. The "time to repair" and "total closures repairable" are the two primary inputs for recovering from runway closures. The next four columns give the number of TBM and cruise missile defenses for each layer of the air defense system. Not shown here, but also included in the cruise missile defenses, would be the number of DCA combat air patrols (CAPs) protecting each location. In this example, the three CSG locations keep a single two-ship airborne, and Base 1 and Base 4 have two 2-ships airborne each.

Table 5.3. Baseline Resilience Measures at Each Location

	Shelters	Time to repair runway closure (hrs)	Total closures repairable	# TBM outer layer sites	# TBM inner layer sites	# CM outer sites	# CM inner sites
CSG 1	0	48	1	2	2	1	4
CSG 2	0	48	1	2	2	1	4
CSG 3	0	48	1	2	2	1	4
Base 1	0	8	2	0.5	0	1	2
Base 2	0	8	2	0.5	0	1	0
Base 3	0	8	2	0	0	2	0
Base 4	15	8	2	1	2	2	2
Base 5	61	8	2	0	0	1	0
Base 6	0	8	2	0	0	1	0
Base 7	0	8	2	0	0	0	0
Base 8	0	8	2	0	0	0	0

As can be seen, we have a selection of close-in and more-distant bases, as well as some that are heavily defended and others less so. In general, the closer bases are better defended. The carrier strike groups (CSGs) and Bases 1 and 4 are also set to use some of their air-to-air aircraft in the cruise missile defense role. Initial sortie rates range from around 1.0 for the more distant bases to 2.0 for the closest ones. Note that CSG locations have somewhat lower sortie rates than land bases because we assume 18 hours of operation as opposed to 24.

Figures 5.1 and 5.2 show the daily sortie generation capability for the aircraft and locations before the conflict begins. There are a total of 660 sorties per day flown by the 564 aircraft. This is somewhat less than that expected by simply taking the average sortie rate times the number of aircraft, because of the sorties lost to cruise missile defense at the five locations using them (the three CSGs and Bases 1 and 4). The fuel storage, offload capacity, and the 108 deployed tankers are sufficient to support these aircraft, so there is no sortie rate degrade from lack of fuel prior to the conflict. Since this notional example contains no warfighting context, such as objectives to achieve, we will simply refer to this starting sortie generation capacity as our reference point or goal and compare sortie generation capable with these starting values.

Figure 5.1. Sortie Capability by Aircraft Type, Prior to Attacks

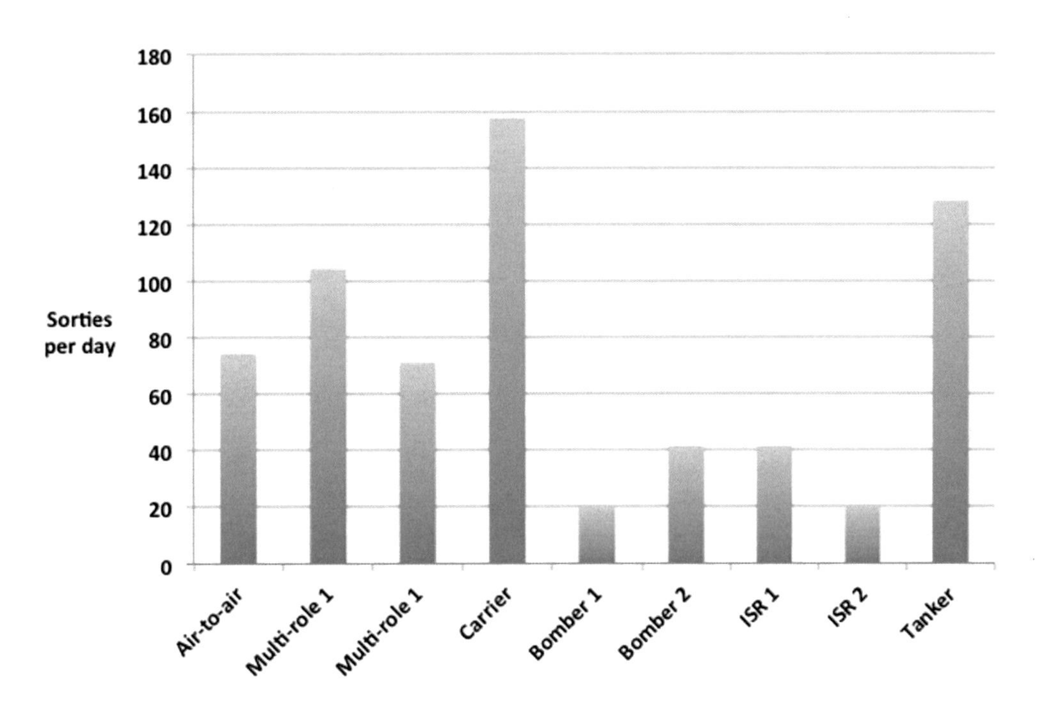

Figure 5.2. Sortie Capability by Location, Prior to Attacks

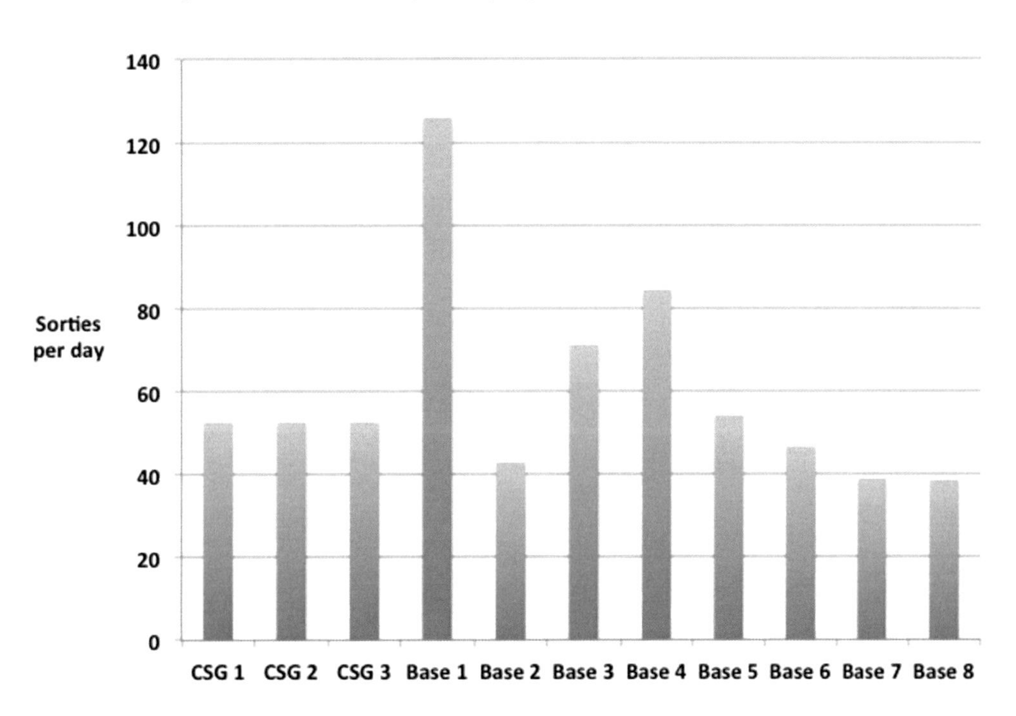

In this simple example, Red's Day 1 ballistic missile attack strategy concentrates on attacking air defenses where they are present, and runways and parking areas otherwise. If we advance the model to the end of Day 1 (we are using 12-hour time steps in this example), we can examine the effect of a single set of missile salvos (90 TBMs and 100 cruise missiles fired) on sortie generation. These results are shown in Figures 5.3 and 5.4 as additional bars added to those in Figures 5.1 and 5.2.

Figure 5.3. Sortie Capability by Aircraft Type, Prior to and After Initial Attack (End of Day 1)

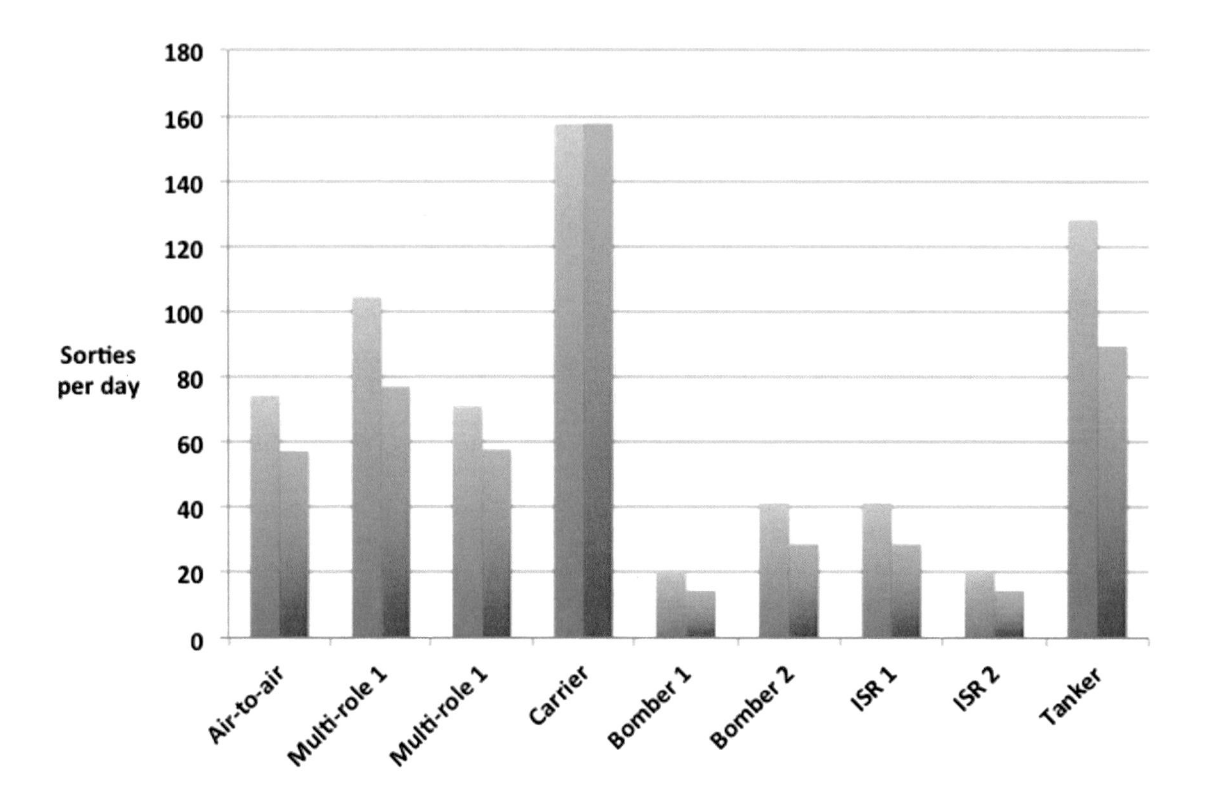

Figure 5.4. Sortie Capability by Location, Prior to and After Initial Attack (End of Day 1)

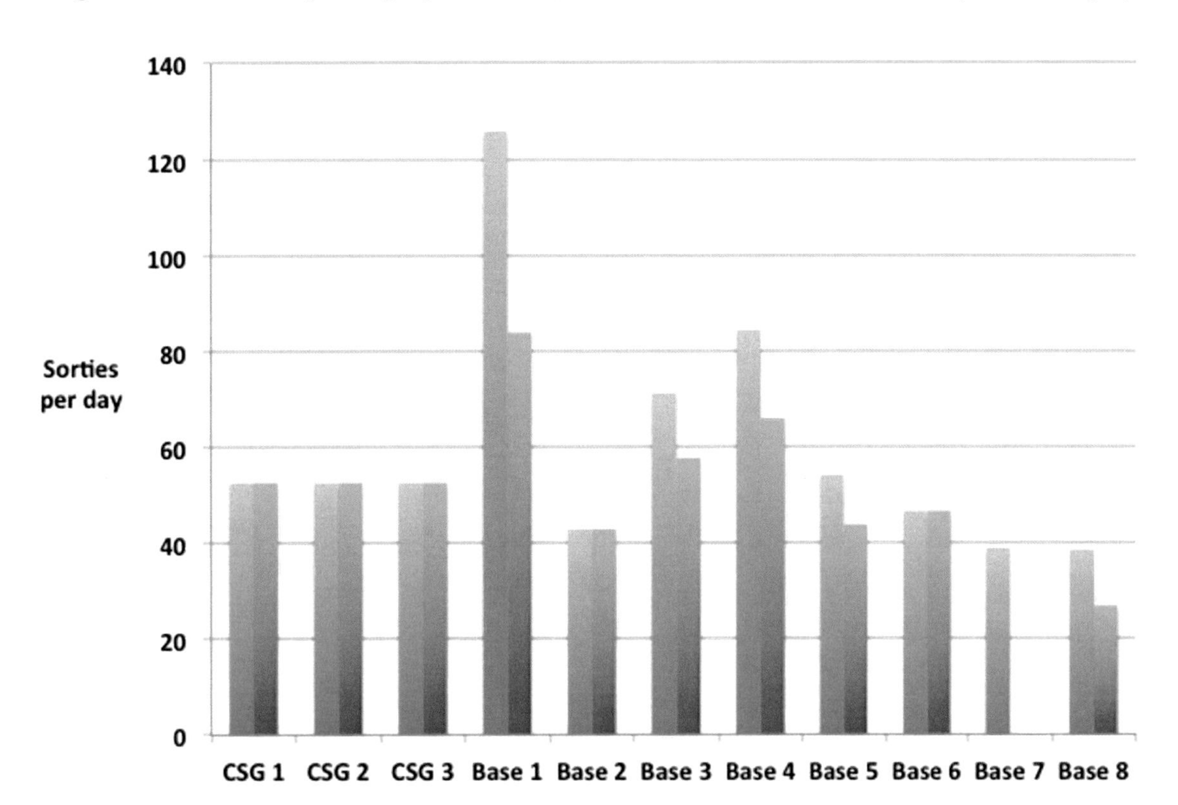

There are several interesting dynamics that can be explored here, but we will investigate two of the more obvious ones: Why is there no degrade to sorties on the aircraft carriers or Bases 2 and 6, and why is Base 7 not producing any sorties at all? The carriers and the two bases, like most of the other locations, were fired at by ten TBMs and ten cruise missiles. Recall that the allocation of missiles against basing targets was heavily focused on air defenses in this initial phase. Figure 5.1 indicates that all three CSG and the two bases have air defenses present; thus, the 20 missiles fired at each location were all targeted at defenses instead of sortie generation–related targets. Indeed, Bases 2 and 6 lost almost all their missile defenses in this attack, although the three CSGs, which were much more heavily protected, were able to defend against this attack with no losses (although almost 100 interceptors of various types were used). The lack of any sorties from Base 7 illustrates the opposite case. Figure 5.4 reflects the fact that Base 7 was left undefended (see Table 5.3); thus, the ten cruise missiles fired at it were targeted at fuel storage. These were sufficient to kill the two storage tanks there; thus, the aircraft at the base were not able to fuel and takeoff, which meant there were no sorties this time step from this base.

On Days 2 and 3, the Red allocation of missiles shifts to a mixed strategy of attacks on defenses if present, as well as runways, parking, and fuel. In the following days, the attacks are concentrated on these latter three target sets, with an equal fraction of attacks on each. The strategy for cruise missiles is similar, except instead of runways and parking areas, aircraft shelters and fuel are substituted until the final phase, when runways are attacked as well. Cruise missiles are used against air defenses in a similar fashion to ballistic missiles.

The firing strategy per day is kept quite simple. Every 24 hours for the first five days, ten ballistic and ten cruise missiles are fired at each base. The exceptions to this are Base 7, which will be designated as only in range of cruise missiles, and Base 8, which is deemed outside the range of all missile types. This attack results in 450 ballistic and 500 cruise missiles being fired at the 11 basing locations over five days.

If we run the model for a full ten days, we can examine the operational MOEs to see how overall Blue capability evolves over time. Figure 5.5 plots the four MOEs as a function of time. As can be seen, the five days of consecutive attacks take a significant toll on Blue's operational capability. On Day 2, the total sorties have dropped from 660 per day to 250, mainly because only 30 percent of the needed aerial refueling capability is available. Six tankers have been killed at Bases 2 and 6, but more importantly all three tanker bases had their runways closed for eight hours. Furthermore, Base 7, which houses one-third of the tanker fleet, had its fuel storage totally destroyed and so cannot fly any tanker sorties at all until repairs begin (set to 10 percent per day in this example). In addition to the tanker effect, five of our 11 runways are at least partially closed, two of our bases have lost more than half their fuel storage, and 18 aircraft have been killed on the ground. These issues are less important, since there would not be sufficient tankers to allow these sorties to fly anyway. The degrade to the aerial refueling capability is particularly devastating because it affects every other sortie that is attempting to fly. Even long-range strike sorties, which need fewer refueling stops, are still limited if the orbits they do require do not have enough tankers on-station to support them.

Figure 5.5. Operational MOEs Over Time

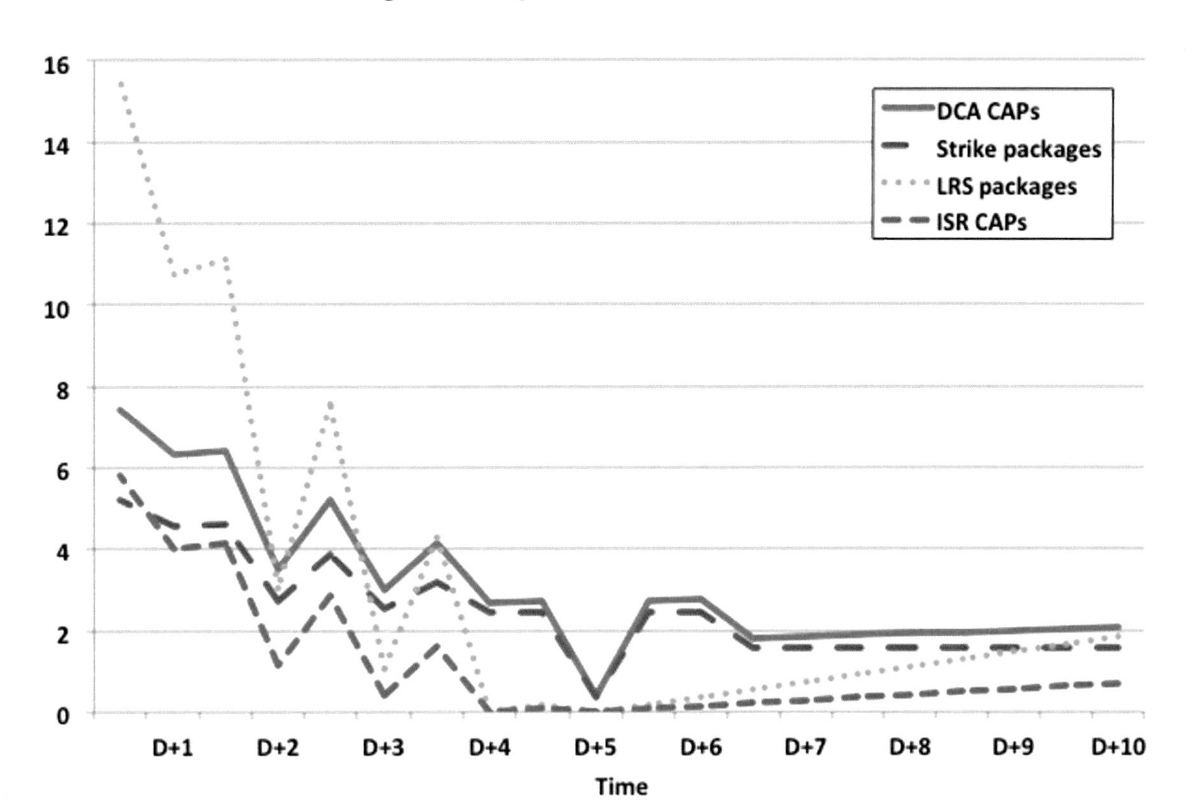

As before, we can examine some of the most interesting dynamics in a bit more detail. For example, why the oscillation in the MOEs from time step to time step? Why are there almost no sorties flown on D+5? The up-and-down behavior of the MOEs is caused by the 12-hour time step in conjunction with the 24-hour missile attack interval. Thus, every other time step is free from missile attack, and so runways will be open (recall that the longest runway repair time is set to eight hours, so runways always reopen before the next time step) and some repair of fuel storage and offload will have occurred. On Day 5, which is the last day of Red's attacks, all the land bases are closed, and the few remaining sorties flown are from carrier air. The land bases are not functioning because five of the eight have exhausted their runway repair materials and, thus, will remain closed for the remainder of the conflict. Unfortunately for Blue, these five include all the tanker bases, which means there is no aerial refueling available until the tankers are relocated. Carrier air, because of its closer basing and relatively long-range aircraft in our example, is the only type that does not require aerial refueling. However, even these sorties are severely degraded, because all the CSG missile-defense interceptors were depleted and the carriers themselves were damaged for the first time, resulting in only 25 percent of the original sortie rate.

Beyond Day 5, the sorties gradually increase. This is mainly because of the repair of fuel storage on Base 7, which is back up to 50 percent of capacity by Day 10; thus, refueling tanker capacity is also slowly improving. The long-range strike sorties improve faster than the other MOEs because they rely less on refueling capacity.

Blue Resilience Improvements

The results shown in Figure 5.5 do not appear particularly positive and could easily be insufficient to support Blue's campaign objectives. So, how might Blue improve its resilience in our simple notional case? To find out, we add three different measures, run the model again, and compare our MOEs with those seen in the figure.[4] First, we triple runway repair capacity at all land bases from the ability to repair two runway cuts to six. This may not be a particularly expensive measure to implement, because most of what is required is raw materials, such as gravel and concrete. Second, we add a fourth tanker location with an additional 20 tankers at Base 4, because it is well defended and not full to capacity. We cannot add 36 tankers here, as we have at the other locations, because the fuel offload capacity at Base 4 will not support them.[5] And third, we add two missile defense units to Base 7 (one TBM inner and one cruise missile outer) to help defend this critical tanker location.

As Figure 5.6 illustrates, these changes roughly double or triple the capacity of Blue, at least as measured by the four metrics. There are now four DCA CAPs during the last time step versus two in Figure 5.5, almost three strike packages per step versus under two previously, seven long-range strike packages now versus two previously, and more than two ISR orbits sustained versus less than one prior to the improvements. We still see very few sorties flown on Day 5. This occurs because of large naval air degrades from the carrier attack and aerial refueling being degraded from loss of fuel on Bases 2 and 7, but those bases recover on subsequent days (as opposed to earlier, when the exhaustion of runway repair materials meant the base was closed permanently), and tankers at Bases 4 and 6 continue to fly. The additional air defenses keep the bases operating at a higher level for longer, and the additional runway repair materials keep the bases operating over the longer term, while the additional tanker location provides redundancy.

[4] Obviously, a more thorough analysis would examine a much broader trade space of resilience measures and parameterize many of the capacity variables.

[5] The fuel storage at Base 4, set to a notional value of almost 10 million gallons, is certainly sufficient, but the offload is 600 gpm. The capacity, which is equivalent to 240,336 lbs/hr, allows just under 40 tanker sorties per day to operate. With a sortie rate of 1.3 at Base 4, 36 tankers would fly almost 47 sorties. Thus 20 tankers at Base 4 is a bit conservative, but does allow for some spare capacity in case of further damage.

Figure 5.6. Operational MOEs with Resilience Improvements

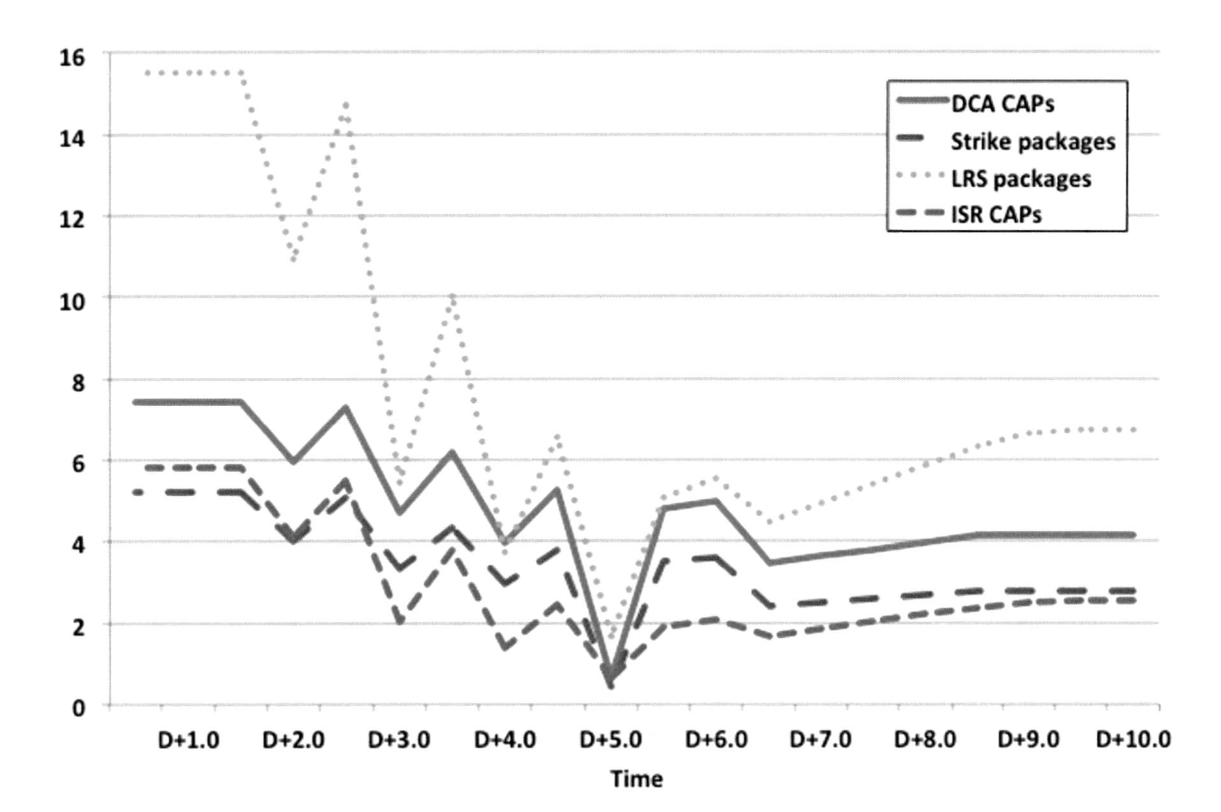

Red Response to Blue's Improvements

Of course, real-world conflicts are a two-sided game, which raises the question of how Red might react to Blue's additional resilience measures. One approach to dealing with defenses is to attempt to overload them with missiles, so what if Red consolidated its five current attacks into two larger ones? This may require an investment in additional launch vehicles, but this could be cheaper than buying additional missiles. Another approach that may be effective for Red is to avoid targeting runways at all, given that Blue is more heavily invested in repair, and instead attack fuel and parking areas to create longer-lasting effects. Finally, Red could decide that focusing attacks on a smaller number of critical bases, namely the four tanker locations, would be more effective than attempting attacks against all 11 sites.

Figure 5.7 once again plots the four MOEs for this case with Red's response. Comparing it with Figure 5.6, there is clearly a drop in Blue's effectiveness, but that drop in effectiveness does not go down to the original levels seen in Figure 5.5. Blue's set of resilience measures appears robust to this particular Red strategy change. The initial set of attacks on Day 1 show some degrade to Blue, but not as large as what we saw at the second attack on Day 3. This is primarily because of the fairly significant allocation of missiles against air defenses in the first attack. However, the pure allocation against tanker bases results in over 40 tankers being lost on Day 1, primarily at Bases 2 and 6, which see over 45 missiles attack their parking areas for essentially complete coverage. However, because of the additional 20 tankers added by Blue, approximately 80 percent of the tanker demand is still met on Day 1, since there was excess capacity at the beginning of the conflict.

Figure 5.7. Operational MOEs with Red's Response to Blue's Resilience Improvements

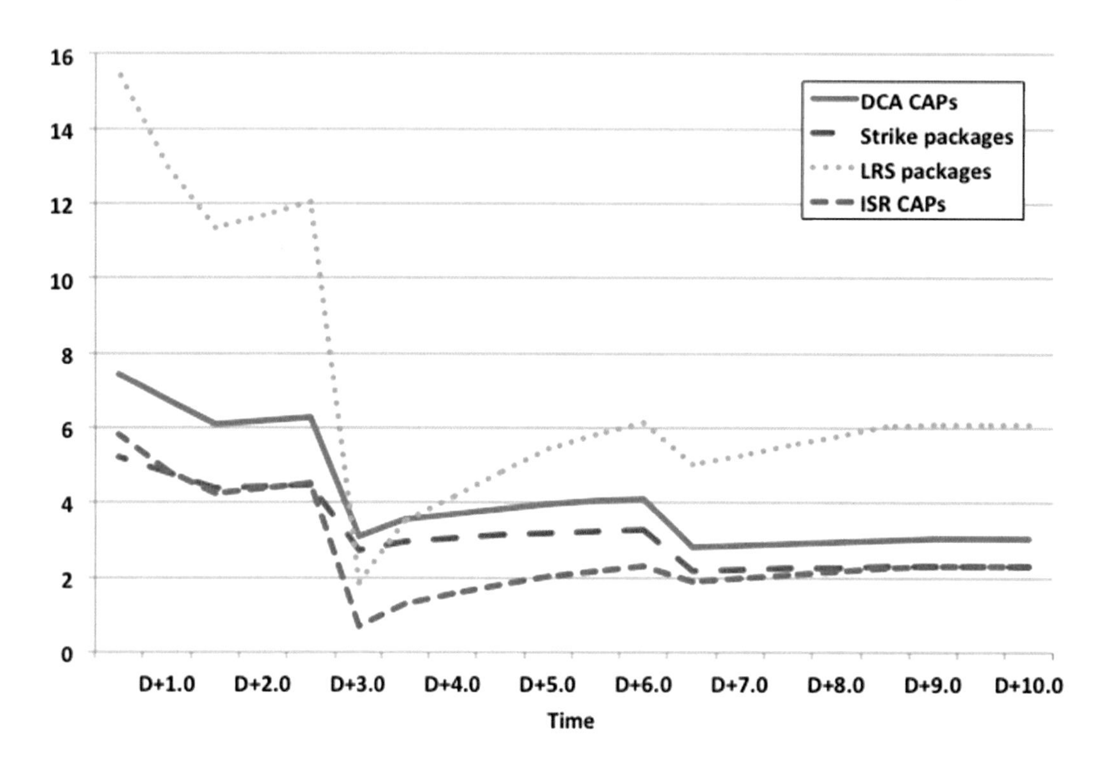

In the Day 3 attack on the tanker bases, another 28 tankers are lost on the ground, and the attacks on fuel storage allow only Base 4 to continue to produce tanker sorties on this time step. Only 15 percent of the tanker requirement can be met, which accounts for the very low number of total sorties produced. As fuel storage is replaced at 10 percent of capacity per day, the sorties continue to climb each day until reaching the levels seen in Figure 5.7. By Day 10, approximately 60 percent of the tanker requirement is being met, and the four tanker bases all have at least 70 percent of their fuel storage operational. At this point, the main constraint is the lack of tanker airframes because of the losses on the ground, which total 68 of the original 128.

We can compare the overall sortie generation of the these three cases in Figure 5.8, which plots the Blue operational sorties (those going toward DCA, strike, LRS and ISR missions) over the ten-day campaign. The baseline case has the lowest sortie total at each time step, with the case with additional Blue resilience improvements showing the highest sortie totals. At many points there are twice as many sorties being generated by this improved Blue case, and the total sorties are 63 percent higher (3,309 versus 2,026). As expected, the change in Red's strategy, plotted as the green dashed line in the figure, falls between these two extremes. Blue still sees an improvement in sortie generation, but the gain is smaller than was seen against the baseline Red attack strategy. However, 42 percent more sorties overall were flown (2,882 versus 2,026), which is a significant and noteworthy improvement in capability for Blue. As noted earlier, with no larger campaign context to refer to in this notional example, it is difficult to determine how many sorties are sufficient. However, we can note that our case with Blue improvements and a Red reaction is generating less than half the starting sorties per 12-hour time step. This implies that, in A2 environments, initial force deployments must be scaled toward significant capacity decreases in the first several days of a conflict and that warfighting objectives must at least consider that much less airpower may be generated than is theoretically possible.

Figure 5.8. Operational Sorties Compared

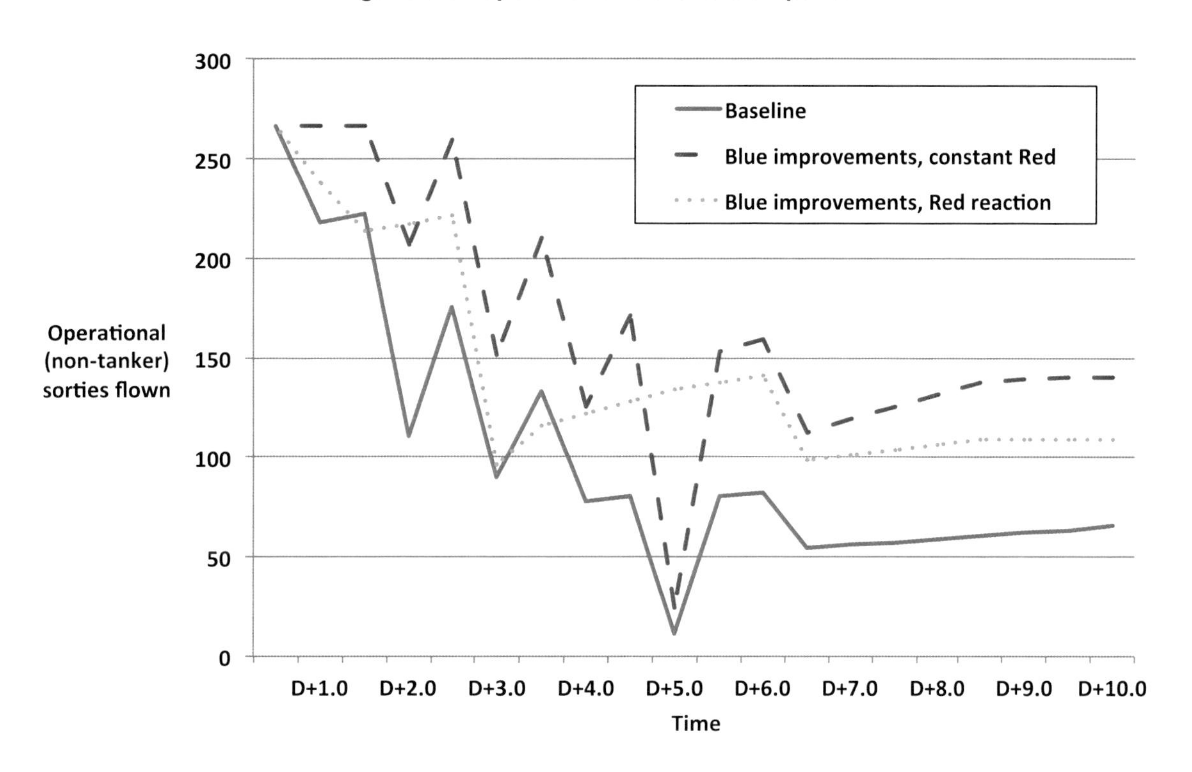

6. Potential for Follow-On Work

Although the work performed in this study was intended to allow both broad and deep analysis of a complex set of resilience issues, there are several areas in which further follow-on work could be done, some of which have already been highlighted in the main text: improving the functionality of ORAM, addressing data uncertainty in the model, expanding the interactive analytical framework in which the model is employed, and automating a greater portion of the analytical framework in ORAM.

Improving the Functionality of ORAM

The largest and likely most important way in which ORAM's functionality could be improved is to add the ability to look at the effect of damage to the logistics and supply chain supporting each airbase. Supporting equipment and personnel, such as ports, road and railways, maintenance hangers, spares, test equipment, munitions, and food, water, and health care for maintainers and pilots, are likely critical to sortie generation, and yet there does not appear to have been recent analyses of their susceptibility to attack or the effects of their loss or damage.[1] ORAM does not yet include these effects in the model, although other work at RAND is currently analyzing what inputs and what levels of degrade would be important to consider. The outputs of this effort may be incorporated into ORAM when that analysis is complete. In addition, some resilience concepts rely on airbases providing mutual support to each other in case of attack.[2] Since most of this mutual support would be in the areas just highlighted, it will be difficult to analyze these concepts without better understanding of the logistics system, both locally at bases and in region when they come under attack. Ongoing work at RAND in FY 2015 is focused on some of these issues, particularly the vulnerability of logistics supply chains and on the effects of personnel losses and injuries.

One can also imagine actions that Red forces could take that might interrupt the resupply of fuel to airbases. Similarly, a successful strike against the pumping infrastructure of the airbase could potentially degrade the base's ability to replenish its fuel stores. As mentioned in Chapter Three, the Red government could also attempt to use political pressure to keep a host country from allowing its refineries to supply American bases with fuel. While these types of attacks were not included in our current modeling effort, further work examining logistics and supply

---

[1] There was substantial research into some of these areas during the Cold War; for example, see RAND's work on the TSAR and TSARINA models documented in Emerson, 1982.

[2] Forthcoming work by Lostumbo et al. (not publicly releasable) explores a concept known as "cluster basing," in which several locations within 100 to 200 miles of each other can benefit from sharing missile defenses and providing logistical support to each other as individual locations come under attack.

chains would likely have to address them. ORAM includes sufficient "hooks" and interrelationships that such explorations could quite easily be done.

Other ORAM functionality issues that likely deserve further attention across these types of aircraft carrier and airbase infrastructure categories are the repair capabilities and capacities. How long would holes in a carrier flight deck take to repair? What are repair times for aviation fuel storage tanks and offload hydrant systems? Could parking areas be easily repaired with flexible mats? Are carrier catapult and landing systems uniquely vulnerable to damage? Would attacking personnel in barracks or tent cities have a catastrophic effect on all other repair types? These questions and many more seem to have received little recent attention by the military services and the analytic community and yet may have a very large effect on accurately measuring airbase resilience. As above, some exploratory analysis would be useful to determine thresholds at which these issues may become critical.

In this area, the potential deployment of temporary fuel bladder systems with sophisticated pumping equipment is a good example. The use of fuel bladders may be a useful resilience measure and allow the use of airbases otherwise ill-equipped for military operations. These systems could possibly be deployed on some of the more remote basing options, especially in the Philippines. Deployable aircraft shelters or decoy aircraft could be similarly interesting resilience measures.

Yet another model functionality improvement has to do with defending against cruise missiles. The inputs of altitude, radar cross-section, and speed are used to determine the maximum engagement range and the number of subsequent salvos of cruise missile defenses. One issue that may be useful to capture in follow-on work is the effect of forward air defenses against the aircraft launching cruise missiles. Killing them, as opposed to each individual cruise missile, is likely to be a much more efficient use of fighter aircraft. The utility of degrading the CEP of various missile types with jamming would be interesting to explore as well.

Another area has to do with calculating refueling orbits. Once the orbits have been located and matched with each location, the demand and supply and any required limits can be calculated for the current time step. We first calculate fuel storage and offload capacity needed at each base to fill each sortie with fuel; this calculation is used to ensure that enough fuel handling is available to launch the time step's sorties, as discussed earlier. Next, the total airborne demands of each location are calculated using the distance, fuel flow, and any loiter time for each type of aircraft. However, we do not include tankers in this calculation. This implies that tankers are not allowed to refuel other tankers, which may be an interesting capability to explore in further work.

Addressing Data Uncertainties with ORAM

Many of the limitations with using ORAM to conduct broad analyses of basing resilience are the result of uncertain or missing data on real-world effectiveness as opposed to a lack of included functionality. One key example of this is the performance of ballistic and cruise missile defenses, which have never been tested against the large raids with countermeasures expected in many scenarios. Although some types of missile defenses are likely to have low effectiveness simply because of physical limitations against countermeasures, even cruise missile defense by fighter aircraft, which may be the most straightforward approach, has seen little real-world testing. Although definitive answers to real-world performance are unlikely to obtained only through testing, the uncertainty bands could likely be narrowed from large unknowns that currently prevail.

Other areas with large uncertainty include many types of repair times and repair capacity, infrastructure operability at less well-known locations, and some types of information on Red capabilities, such as longer-term ballistic and cruise missile operations and warhead availability. Although reasonable estimates can be made for many of these parameters, intelligence collection, testing, and analysis improvements will need to be made to provide robust comparisons between possible resilience measures. As with other variables, parametric excursions across these uncertainty bands would help highlight the criticality of these inputs.[3]

For example, as noted above, the input for fuel storage repair is currently simply set to 10 percent per day. This means that if 20 percent of the fuel storage were destroyed on Day 1, it would be back to 100 percent at the beginning of Day 3. Obtaining more realistic information on fuel storage repair rates, especially at various types of bases, would be an important area to explore in further work.

As with fuel storage, more research is also needed in dealing with fuel offload repairs. As noted above, fuel offload repairs have a repair rate per day, currently set to 25 percent to reflect a presumed greater ease of repair (and possibly greater scope for workarounds) compared with fuel storage tanks. However, this may not always be the case—for example, if fuel tanks are easily patched but fuel hydrant systems are hardened or underground and rely on valves and pumps that are not easily replaced.

Even if all of these areas of highly uncertain information were addressed, however, many others will remain.[4] As a result, any broadly useful analysis of basing resilience must embrace techniques to deal with high levels of uncertainty in inputs and Red and Blue strategy. Automating the running of the tool to accommodate ranges of input variables, instead of point estimates, has been done in limited cases but could be greatly expanded. As discussed earlier, creating multiresolution subcomponents of the model would be helpful in focusing analytic and computing resources where they are most needed.

---

[3] Recent internally funded RAND work by Scott Grossman and other have explored the use of fuzzy logic algorithms for handling the large uncertainty bands of many of these variables. Although ORAM is likely unsuited for this type of analysis, if the uncertainties cannot be reduced, these types of techniques may be required.

[4] And more to the point, even "definitive" sources of data should be cautiously treated as uncertain.

Expanding the Game-Theoretic Analytical Framework

The game-theoretic framework used in this research allowed us to explore the interactive nature of the competition between A2/AD developments and operational resilience investments; however, the simplified way in which we applied it introduced a couple of artificialities that can be avoided in future analyses. First, setting each case run for a specific conflict length, known to both sides in advance, created incentives for each side to develop strategies that might diverge from those employed in actual conflicts. Knowing that the notional scenario would end in a specified number of days encouraged analysts to expend the entire Red missile quiver in that time period; however, in an actual conflict, analysts would have to hold some number of missiles in reserve to hedge against uncertainty. Similarly, analysts were inclined to hold Blue assets out of range of Red missiles until the Red quiver was nearly depleted, knowing Red would need to expend its missiles before the end of the notional scenario.

Second, although we have occasionally referred to the Red targeting plans we experimented with as "strategies," the interactive analysis we conducted was actually strategy-agnostic. No consideration was given to what Red's objectives might be, what campaign plan it might develop to obtain those objectives, and what targeting it would need to best support that campaign. These elements could vary widely from one conflict to another—an adversary's decision about whether to conduct a maritime blockade, coercive bombardment, or amphibious invasion would result in campaign plans that differ widely in terms of objectives, phasing, and expected duration. Those factors, in turn, would call for very different Red missile targeting plans and rates of expenditure.

Rectifying these deficiencies can enhance the utility of future operational resilience analyses using the interactive framework. Instead of executing generic targeting plans in conflicts of fixed duration, researchers should construct scenarios simulating a broad range of estimates of what a potential adversary's campaign plans would resemble and develop Red's initial missile-targeting plans based on those scenarios. Interactive analysis could proceed from there, exploring conflict durations based on estimates of Red's expected conflict objectives to measure the effects on each side's strategies.

Broader Uncertainty and Sensitivity Analysis

Although our work to date has highlighted a few important uncertainties, analysis with ORAM to date has not explored many other critical scenario variables. A more extensive uncertainty analysis may generate considerably many additional concerns about airbase resilience in some respects and less in others. A key priority is the ability to assure robustness, hard-headedness, and comprehensibility of uncertainty-sensitive analysis. While the interactive analytical framework we employed in this study relies on the analyst to examine and manually adjust Red and Blue strategies at each step, much of this iterative process could be automated. Automation would enable consideration of a wider range of strategic options. Automation could also be married with an optimization algorithm to find the most cost-effective resilience improvements in the face of an adaptive Red force.[5] Resources and time constraints did not allow us to make these extensions in this project, because our first concern was to explore and illustrate the strategic dynamics of the operational resilience challenge. Requiring analysts to manually explore and manage the process of developing Blue and Red competing strategies has been very useful to understand the strategic context of the problems raised by A2 threats and basing resilience. Given how important assumptions about Red attack strategies are and how complex the interaction between a mutually adaptive Red and Blue can be, automated strategy development can run the risk of obscuring key uncertainties rather than highlighting them. However, proper exploration of the large parameter space inherent to resilience problems likely requires at an automated "sweep" function to ask the big "what if" questions, such as what if repair rates were faster/slower, consumables were more/less available, missile defenses were more/less capable, etc. This approach, as well as optimizing algorithms for Red and Blue, should be explored in future research.

Redesign to a Multiresolution Version of ORAM

As discussed throughout this report, ORAM contains quite a bit of detail for many areas of analysis. A more elegant approach would be to re-create the model (or as a family of models) with more detail where it is required for correct results and less where possible to better enable broad searching of parametric input spaces without distorting the outcomes.[6] This approach can enable an analyst to "have one's cake and eat it too," but would require redesign, trade-offs between speed and detail, and investment into recoding. To make best use of this investment, such work should include close cooperation with other resilience modeling efforts, such as the RAND CODE team, since decisions would have to be made about the relative role of ORAM and CODE, as well as the relationship between them.

---

[5] Conducting such an analysis would be an application of portfolio analysis, which is an analytic approach that helps decisionmakers in balancing diverse investments to meet objectives while balancing risk. A familiar application is managing a financial portfolio of stocks and bonds to maximize expected returns at a given level of risk. See Markowitz, 1952, for a classic treatise on the subject. More general applications use related concepts to maximize mission performance for a given budget constraint. As the number of potential investments that could go into a portfolio increases, the set of potential portfolios that could be made of those investments increases exponentially. Formal portfolio analysis uses operations research methods, such as mixed integer programming, to identify the best mix of investments, that is, the top-performing portfolios affordable within different budget limits. As with all optimizations, the selection of the objective function (i.e., how one defines one's goal) can have profound consequences for the relevance of the optimization's results. For a more strategic view of portfolio analysis less reliant on mathematical manipulation, see Davis, Shaver, and Beck, 2008. There are also tools available to support this type of portfolio analysis, for example see Davis and Dreyer, 2009, and Moynihan, 2005.

[6] For discussion of multiresolution modeling with several examples relevant to national security analysis, see Davis and Bigelow, 1998.

Appendix A. Operational Resilience Lexicon

One of the tasks of this project required developing an operational resilience lexicon for evaluating the potential impacts of alternative attacks on U.S. air forces, theater-wide, and the potential benefits of a wide range of approaches for improving operational resilience, singularly and in combination, in the face of those attacks. This list of terms and definitions drew heavily on the *Department of Defense Dictionary of Military and Associated Terms* (JP 1-02, 2014), because the definitions provided there have been coordinated with and agreed upon by the military services and other agencies within the defense community. However a number of concepts central to operational resilience are not addressed in the dictionary, and those that do appear are often defined at too general a level to be analytically useful. Therefore, working in concert with the sponsor, other interested offices within Headquarters Air Force, and offices at Headquarters Pacific Air Forces, we developed additional definitions to describe various resilience approaches and added granularity to existing definitions where needed to make them more useful as metrics for analysis. In the list of terms below, we have indicated in brackets whether a definition comes primarily from JP 1-02, some other source, or was developed as part of this project.

Access. Characteristics of a country, base, operating location, or facility that determine the degree of freedom in which U.S. forces will be allowed to operate there in times of peace or war. [RAND]

There are two main categories of access: base access and country overflight:

- **Base Access** is categorized into four levels:
  - *Level 1:* **Limited access.** Blue force aircraft are not allowed to base there. They may occasionally land there in times of peace, but might not be allowed access in times of war. No combat or combat support operations may be flown from or recovered at a Level 1 base. [RAND]
  - *Level 2:* **Restricted conditional access.** Blue force aircraft may base there subject to conditions set by the host government regarding types of aircraft or operations that may be associated with that base. Combat missions may not be flown from a Level 2 base without explicit host approval; however, standing approval may be granted for some combat support missions or emergency recovery of Blue aircraft returning from combat missions. [RAND]
  - *Level 3:* **Broad Conditional access.** Blue force aircraft may base there subject to conditions set by the host government regarding types of aircraft or operations that may be associated with that base. Host government has granted standing approval for certain types of combat and combat support operations. Such approval may apply only to specific scenarios or operations against specific opponents. [RAND]

- Level 4: **Assured access.** Blue force aircraft may base there and conduct or recover combat and combat support operations without constraint or prior approval. Assured access bases are on U.S. territory. [RAND]

- **Country Overflight** is categorized into three levels:
  - *Level 1:* **Limited overflight.** Blue force aircraft may overfly the country's territory for noncombat and noncombat support missions in peacetime, but might not be allowed overflight in times of war. No combat or combat support operations may be flown over Level 1 countries. [RAND]
  - *Level 2:* **Restricted conditional overflight.** Blue force aircraft may overfly the country's territory subject to conditions set by that government regarding types of aircraft or operations that may flown there. Level 2 countries typically do not allow Blue aircraft to overfly their territories on combat or combat support missions. [RAND]
  - *Level 3:* **Broad conditional overflight.** Blue force aircraft may overfly the country's territory subject to conditions set by that government regarding types of aircraft or operations that may flown there. Host government has granted standing overflight approval for certain types of combat and combat support operations. Such approval may apply only to specific scenarios or operations against specific opponents. [RAND]

Active Defense. The employment of limited offensive action and counterattacks to protect a friendly air, land, sea, or space platform or to deny a contested area or position to the enemy. Active defenses encompass the use of electronic warfare, antiaircraft weapons such as anti-aircraft artillery (AAA) and surface-to-air-missiles (SAMs), fighter interceptors, ground forces, and other means of reducing the effectiveness of incoming attacks. [JP 1-02, RAND modified]

There are two types of active defense against air and missile attacks, each with multiple levels of capability.

- **Air Defenses.** Defensive measures designed to destroy attacking aircraft or cruise missiles or to nullify or reduce the effectiveness of such attacks. [JP 3-01, 2012, RAND modified]

  There are three levels of air defense:

  - *Level 1:* **Basic**. Fighters scrambled on warning from ground bases or aircraft carriers, or surface-based terminal defenses, such as AAA or SAMs. [RAND]
  - *Level 2:* **Postured**. Fighters on CAP vectored to hostile intruders. [RAND]
  - *Level 3:* **Layered**. Fighters on CAP, supported by additional aircraft scrambled as needed, with friendly bases and forces also supported with terminal defenses, such as AAA or SAMs. [RAND]

- **Ballistic Missile Defenses (BMD).** There are three levels of BMD:
  - *Level 1:* **Basic**. Terminal defense only [RAND]
  - *Level 2:* **Postured**. Terminal and mid-course defenses [RAND]
  - *Level 3:* **Layered**. Terminal, mid-course, and boost-phase defenses. [RAND]

Aerial Port Operation. A logistical operation designed to support the functioning of an airfield that has been designated for the air movement of personnel and materiel. [JP 3-17, 2013, as "Aerial Port," RAND modified]

Airbase. A land facility built, equipped, and supplied to project or support some level of military air operations. An airbase is often described as a "land base" or simply a "base." [RAND]

There are ten types of airbases:

- *Type A*: **Main Operating Base (MOB).** A facility outside the United States and U.S. territories with permanently stationed operating forces and robust infrastructure. Main operating bases are characterized by command and control structures, enduring family support facilities, and strengthened force protection measures. [JP 1-02, 2014]
- *Type B*: **Forward Operating Base (FOB).** An airfield used to support tactical operations without establishing full support facilities. The base may be used for an extended period of time. A FOB typically requires some level of support from a MOB. [JP 1-02, 2014]
- *Type C*: **Forward Operating Location (FOL).** A forward operating base that is served by a less extensive support structure than used in an FOB. FOLs are primarily used in counter-drug operations. [JP 1-02, 2014]
- *Type D:* **Forward Support Location.** A supply base designed to support operations at one or more forward locations during a military conflict. [RAND]
- *Type E*: **Tanker Base**. Any base that has the capability of supporting tanker operations and is designated to perform that function. Bases can be considered tanker bases if they support other types of operations as well. [RAND]
- *Type F*: **Dispersal Base.** A base to which combat aircraft can deploy from another base or operating location to effect dispersion. Dispersal bases should have sufficient fuel, ammunition, and support to carry out combat operations for seven days. [JP 1-02, 2014 as "Dispersal Airfield," RAND modified]
- *Type G*: **Recovery Base.** A base with the purpose of repairing, refueling, and resupplying an aircraft after a sortie. [RAND]
- *Type H*: **Supply Base.** Any military installation with a store of supplies large enough to require personnel assigned to maintain them and coordinate their distribution. [RAND]
- *Type I:* **Flex Base.** A base within about 100 miles of the main operating base or dispersal base to which aircraft can divert to refuel then return to the main operating base or dispersal base when possible. These bases must be capable of providing a minimum five days of support to launch, recover, refuel, and rearm aircraft in support of the air tasking order, although perhaps at a reduced level. [Pacific Air Force (PACAF)/A5X, RAND modified]
- *Type J*: **Cluster Base.** One of several bases (conceptually 5–10) to which fighters can disperse, which are close enough to share air and missile defenses and some elements of logistics support. Each cluster base should be capable of providing a minimum of three days of logistics support to launch, recover, refuel, and rearm aircraft in support of the air tasking order. [RAND]

Airbase Defense. Active and passive measures taken to defeat or minimize the effects of attacks on airbases by conventional or unconventional surface forces. [RAND]

There are three levels of airbase defense:

- *Level 1:* Defenses against enemy agents, saboteurs, sympathizers, terrorists, and civil disturbances. Typically requires basic physical security measures only, such as access control and routine policing, but may require increased security at terrorist force protection conditions Bravo or higher. [RAND; Air Force Doctrine Document [AFDD] 3-10, 2011]
- *Level 2:* Defenses against small tactical units, unconventional warfare forces, guerrillas, and may include significant stand-off weapons threats such as those imposed by mortars, rockets, rocket propelled grenades, and SAMs. Level 2 defenses typically entail a robust base security force and may include passive defenses such as hardening and dispersal and support from friendly forces outside the base boundary. See "Passive Defenses." [RAND; AFDD 3-10, 2011]
- *Level 3:* Defenses against large tactical force operations, including airborne, heliborne, amphibious, infiltration, and major air operations. Level 2 defenses include a robust base security force, passive defenses, and support from friendly forces outside the base boundary. [RAND; AFDD 3-10, 2011]

Air Superiority. That degree of dominance in the air battle by one force that permits the conduct of its operations at a given time and place without prohibitive interference from air and missile threats. [JP 3-30, 2014]

The degree of air superiority attainable can be described in three levels:

- *Level 1:* **Contested Airspace.** Necessitates continuous combat operations to maintain sufficient control of an area. [RAND]
- *Level 2:* **Controlled Airspace.** Requires only intermittent sorties to neutralize Red anti-air actions, such as when an operational area is located near enough to a friendly airbase that Blue has time to obtain and act upon early warning information and defeat Red attempts at incursion. [RAND]
- *Level 3:* **Air Supremacy.** That degree of air superiority wherein the opposing force is incapable of effective interference within the operational area using air and missile threats. [JP 1-02, 2014]

Anti-Access/Area Denial (A2/AD). A general classification of threats to U.S. abilities to deploy and employ expeditionary forces. See separate definitions for "anti-access" and "area denial." [Joint Operational Access Concept [JOAC] version 1.0, 2012, RAND modified]

Anti-Access (A2). Military measures taken to keep an opponent's forces from entering a theater of operations. [RAND]

There are four types of A2 measures:

- *Type A*: Air and missile strikes launched from land-based platforms against Blue land- or sea-based forces. [JOAC version 1.0, 2012, RAND]
- *Type B*: Maritime attacks launched from sea-based platforms, such as surface force attacks, submarine attacks, anti-submarine warfare, submarine mining, or air and missile attacks. [RAND]

- *Type C*: Attacks on Blue land forces, airbases, ports, or support facilities by Red conventional land forces or special operations forces, or Red directed or inspired irregular forces such as insurgents or terrorists. [RAND]
- *Type D*: Kinetic or nonkinetic attacks on Blue space or cyber capabilities. [RAND]

Area Denial (AD). Military measures designed to keep an opponent's forces from entering a defended area or operating effectively within a theater of operations. There are two types of AD measures: [RAND]

- *Type A*: **Kinetic defenses.** SAMs, interceptor aircraft, coastal patrol craft, mines, attack submarines, etc. [RAND]
- *Type B*: **Nonkinetic defenses.** Radar jamming and other forms of electronic warfare. [RAND]

Asymmetric. In military operations, the application of dissimilar strategies, tactics, capabilities, and methods to circumvent or negate an opponent's strengths while exploiting his weaknesses. [JP 3-15.1, 2012]

Battle Network. All assets necessary to carry out a broadly defined mission (e.g. anti-access and area denial) and the interconnections between them. The battle network–counter battle network competition is designated one of the critical aspects of the AirSea battle concept.

Operational resilience requires the maintenance of the battle network to a level needed to withstand attack, adapt, and support the generation sufficient combat power to achieve campaign objectives in the face of continued adaptive enemy action. [Center for Strategic and Budgetary Assessments, *AirSea Battle: A Point-of-Departure Operational Concept*, 2010, RAND modified]

Camouflage, Concealment, and Deception (CC&D). Any of various types of disguise or actions intended to confuse enemy targeting. [JP 1-02, 2014, RAND modified]

There are three levels of CC&D:

- *Level 1:* **Basic.** Netting, closed hangars, paint, obscurants, etc. [RAND]
- *Level 2:* **Intermediate.** Decoy aircraft, vehicles, structures, etc. on existing bases. [RAND]
- *Level 3:* **Advanced.** Decoy bases with level 2 decoys, feint deployments and maneuvers, deceptive radio transmissions, etc. [RAND]

Cold Start. The initiation of an expeditionary operation without a preliminary force buildup in the conflict area. [RAND]

Command and Control (C2). The exercise of authority and direction by a properly designated commander over assigned and attached forces in the accomplishment of the mission. C2 functions are performed through an arrangement of personnel, equipment, communications, facilities, and procedures employed by a commander in planning, directing, coordinating, and controlling forces and operations in the accomplishment of the mission. The term also is used to describe the personnel, facilities, and ground, sea, air, space, and cyber platforms used to perform this function. This includes platforms in garrison and ports and airborne assets on the ground at the time of attack. [JP 1-02, 2014, RAND modified]

There are four types of C2 elements:

- *Type A*: **Land-based.** Personnel and equipment operating in land-based facilities. [RAND]
- *Type B*: **Seaborne.** Personnel and equipment operating aboard ships. [RAND]
- *Type C*: **Airborne.** Personnel and equipment operating in aircraft. [RAND]
- *Type D*: **Space and cyber.** Equipment operating in space and virtual structures in the cyber domain. [RAND]

Decoy. An imitation in any sense of a person, object, or phenomenon that is intended to deceive enemy surveillance devices, mislead enemy evaluation, or otherwise increase the number of weapons required to carry out a successful attack or defense (e.g. dummy aircraft, buildings, support vehicles, missile warheads, etc.). [JP 1-02, 2014, RAND modified]

Deployable Shelter. A mobile structure designed to protect its contents from air or missile attack by providing a degree of hardening and concealment. See "Hardening" and "Camouflage, Concealment, and Deception." [RAND]

Denial of Service. Action taken to degrade the performance of an adversary's computer-based network via overloading inputs to one or more key nodes. Examples include mass transmission of messages to Internet servers or sending radio signals to lock uplink receivers on satellites. [JP 1-02, 2014, RAND modified]

Dispersal. Sometimes called "dispersion," the spreading or separating of troops, materiel, establishments, or activities, which are usually concentrated in limited areas, to reduce vulnerability. This concept encompasses both base-level dispersal and theater-level dispersal. [JP 1-02, 2014, RAND modified]

Dispersal Parking. Parking aircraft with large minimum distances between them so as to complicate an attacker's targeting and minimize the risk of chain reactions from the destruction of individual aircraft. This is an example of base-level dispersal. [RAND]

There are three levels of dispersal parking:

- *Level 1:* **Basic dispersal.** Minimum distance of 50 meters between aircraft. This level of dispersal complicates the planning and execution of sapper attacks, such as those from special operations forces, insurgents, or terrorists. [RAND]
- *Level 2:* **Intermediate dispersal.** Between 50 and 120 meters distance between aircraft. This level of dispersal complicates the planning and execution of air and missile attacks. [RAND]

- *Level 3:* **Wide-area dispersal.** More than 120 meters between aircraft. This level of dispersal provides a measure of protection against missile salvos and cluster munitions. [RAND]

Electromagnetic Jamming. The deliberate radiation, reradiation, or reflection of electromagnetic energy for the purpose of preventing or reducing an enemy's effective use of the electromagnetic spectrum, and with the intent of degrading or neutralizing the enemy's combat capability. [JP 1-02, 2014]

Expeditionary. Any power projection operation requiring the use of a forward operating location. Also describes a force that is organized, trained, and equipped to deploy to a forward location and operate quickly. Expeditionary forces include air mobility assets that can deliver forces to locations other than MOBs. [JP 1-02, 2014, RAND modified]

Force Sustainment. The provision of logistics and personnel services required to continue operations until successful mission accomplishment. The force sustainment-counter force sustainment competition is designated one of the critical aspects of the AirSea battle concept. There are three levels of force sustainment. [Center for Strategic and Budgetary Assessments, *AirSea Battle: A Point-of-Departure Operational Concept*, RAND modified]

- *Level 1:* **Bare-bones.** Sustainment sufficient to support one sortie per day for less than the full complement of combat aircraft operating out of the base. [RAND]
- *Level 2:* **Basic.** Sustainment sufficient to support one sortie per day for the full complement of combat aircraft operating out of the base. [RAND]
- *Level 3:* **Robust.** Sustainment sufficient to supply multiple sorties per aircraft per day in intense combat operations. [RAND]

Gaining and Maintaining Access. A joint Army/Marine Corps concept for seizing control of critical areas in the face of anti-access and area denial measures. The operational resilience required for gaining and maintaining access encompasses the ability to withstand enemy attack, adapt, and generate the sorties necessary to insure the success of ground operations in the face of continued, adaptive enemy action. [U.S. Army and U.S. Marine Corps, *Gaining and Maintaining Access, an Army–Marine Corps Concept*, 2010, RAND modified]

Hardening. An element of passive defense involving building or improving structures to protect personnel, equipment, or infrastructure against Red attack. Hardening can be done with permanent structures or deployable shelters (see deployable shelters). [PACAF/A8X; AFMAN 10-2503, 2011; PACAF/Pacific Fleet (PACFLT) Base Resilience Workshop, RAND modified]

There are four types of hardening, which describe sizes and functions of the facilities to be protected:

- *Type A*: **POL/Ammo.** Hardened storage for petroleum, oils, and lubricants (POL) and ammunition. [RAND]

- *Type B*: **ACFT-Small.** Shelters or berms for small aircraft (fighters, small drones, etc.). [RAND]
- *Type C*: **ACFT-Large.** Shelters or berms for large aircraft (tankers, transport, bombers, large drones, etc.) [RAND]
- *Type D*: **Support.** Hardened support facilities (command posts, operations facilities, maintenance shops, barracks, etc.) [RAND]

There are also four levels of hardening, which describe degrees of hardness sought:

- *Level 1:* **Open Shield.** Berms and barricades to complicate planning and execution of sapper attacks and reduce chain-reaction damage from ground or air attacks on individual aircraft or facilities. [RAND]
- *Level 2:* **Basic Shelter.** Hardening against submunition warheads such as Dual-Purpose Conventional Improved Munition (DPICM). [RAND]
- *Level 3:* **Enhanced Shelter.** Hardening against unitary munitions (up to 2,000 lbs). [RAND]
- *Level 4:* **Super-Hardened Shelter.** Hardening against very large conventional or penetrating munitions, usually achieved by deep burial or tunneling. [RAND]

Intense Combat Operations. Operations requiring the generation of the maximum possible number of sorties per day from one or more bases. Often referred to as "surge operations." [RAND]

Integrated Air Defense System (IADS). A system of multiple anti-air sensors and weapons, the operations of which are coordinated by a command and control network, with the purpose of defending a designated area or asset from air attack.

Operational resilience concerns itself with both Blue's ability to defeat Red's IADS—that is, defeat Red's area-denial effort—as well as the Blue IADS which protects friendly forces from Red air attack. [JP 1-02, 2014, RAND modified]

Lethal Area. The area immediately surrounding the point struck by a bomb, missile, or other explosive munition in which unprotected personnel are likely to be killed or unprotected

The lethal area figure applies only to personnel or materials standing in the open. A missile strike against a hardened structure is treated in terms of either total or partial destruction of the structure, along with possible damage to the structure's contents. [JP 1-02, 2014, as "mean area of effectiveness for blast (MAEB)," "mean area of effectiveness for fragments (MAEF)"; RAND]

Assuming an open area, the lethal area will vary, depending on the size and type of the munition. This variance can be divided into three rough levels:

- *Level 1:* **Small diameter.** 10 meters. [RAND]
- *Level 2:* **Medium diameter.** Between 10 and 50 meters. [RAND]
- *Level 3:* **Large diameter.** Greater than 50 meters. [RAND]

Line of Communications (LOC). A route traversing land, water, or air that connects an operating military force with a base of operations and along which supplies and military forces move. [RAND]

There are two types of LOCs:

- *Type 1:* **Essential.** LOCs which, if severed, would result in defeat or mission failure for the military force or base that depends on them. [RAND]
- *Type 2:* **Nonessential.** LOCs which provide useful supplies or forces, but which, if severed, would not result in defeat or mission failure for the military force or base that depends on them. [RAND]

Note: A single LOC can be variously specified essential and nonessential for different operations, e.g., a direct communications link between an airbase and an aircraft carrier might be essential for an AirSea battle operation, but not for a land war.

Maintenance Time. The portion of time during which an aircraft is grounded between sorties that is devoted to repairing damaged or faulty components. [JP 1-02, 2014, as "Maintenance," RAND modified]

Military Conflict. A hostile geopolitical situation in which the belligerents resort to the use of military force to secure their interests. [RAND]

It is useful to describe military conflicts in terms of their nature and approximate duration. For purposes of operational resilience, there are three types of military conflict:

- *Type 1*: **Skirmish.** A brief and often spontaneous combat encounter. All sortie generation takes place in a short time, often over the course of a single day. [RAND]
- *Type 2:* **Operation.** A coordinated set of combat actions carried out to achieve specified military objectives. Sortie generation takes place over several days but usually less than a month. As opposed to a skirmish, which might erupt spontaneously, an air operation involves planning and deliberate execution. [RAND]
- *Level 3:* **Campaign.** A series of related military operations aimed at accomplishing a strategic or operational objective within a given time and space. Campaigns often call for the generation of high volumes of sorties for extended periods of time, sometimes lasting one or more months. [RAND]

Network-Enabled Weapon. A weapon that receives targeting, navigation, or fire-control direction by datalink from one or more remote nodes of the battle network. [RAND]

Operational Access. The ability to project military force into an operational area with sufficient freedom of action to accomplish a mission. [RAND]

Operational resilience can be thought of as the maintenance of operational access in the face of adaptive enemy action.

Operational Resilience. The capacity of a force to withstand attack, adapt, and generate sufficient combat power to achieve campaign objectives in the face of continued, adaptive enemy action. [RAND]

Passive Defenses. Measures taken to reduce the probability of and to minimize the effects of damage caused by hostile action without the intention of taking the initiative. [JP 1-02, 2014]

There are eight types of passive defense:

- *Type A*: **Hardening.** See "Hardening" for types and levels.
- *Type B*: **Dispersal.** See "Dispersal," "Dispersal Base," and "Dispersal Parking" for approaches and levels.
- *Type C*: **Range.** Basing or operating at distances from expected threats that exceed the reach of some and increase response times to those whose reaches are not exceeded. See "Range" for zones.
- *Type D*: **Mobility.** A quality or capability of military forces which permits them to move from place to place while retaining the ability to fulfill their primary mission. (JP 1-02; JP 3-17, 2013)
- *Type E*: **CC&D.** See "Camouflage, Concealment, and Deception (CC&D)" for levels. Note: Level 3 CC&D includes actions that could be classified active defenses.
- *Type F*: **Chemical, Biological, Radiological, Nuclear (CBRN) Defense.** Measures taken to minimize or negate the vulnerabilities and/or effects of a chemical, biological, radiological, or nuclear incident. [JP 1-02, 2014; JP 3-11, 2013]
- *Type G*: **Detection and Warning.** Systems and procedures designed to detect an incoming attack, alert military forces to activate defenses, and warn personnel in the targeted area to take cover or other protective measures. [RAND]
- *Type H*: **Recovery and Reconstitution.** Those actions taken by a military force during or after operational employment to restore its combat capability to full operational readiness. [JP 1-02, 2014; JP 3-35, 2013]
- *Type I*: **Other.** Built-in redundancy, operational security, etc. [RAND]

Power Projection. The ability of a nation to apply all or some of its elements of national power—political, economic, informational, or military—to rapidly and effectively deploy and sustain forces in and from multiple dispersed locations to respond to crises, to contribute to deterrence, and to enhance regional stability. [JP 3-35, 2013]

Precision-Guided Munition (PGM). A guided weapon intended to destroy a point target and minimize collateral damage. Typically, such weapons use a radar, infrared, laser, or electro-optic guidance system or are otherwise guided by an off-board operator. [JP 3-03, 2011, RAND modified]

Range. The distance of a Blue base from the opponent's territory. Blue bases and forces may be postured in any or all of three zones from any given Red territory. [JP 1-02, 2014, RAND modified]

- *Zone 1*: The area between the edge of an opponent's territory and the maximum range of his short-range ballistic missiles (SRBMs), approximately 1,000 km. [RAND]
- *Zone 2*: The area between the edge of Zone 1 and the maximum range of an opponent's medium-range ballistic missiles (MRBMs) and ground-launched cruise missiles (GLCMs), approximately 1,000-2,000 km. [RAND]
- *Zone 3*. The area between the edge of Zone 2 and the maximum range of an opponent's intermediate-range ballistic missiles (IRBMs) and air-launched cruise missiles, approximately 2,000-5,000 km. [RAND]
- *Zone 4*. The areas beyond Zone 3 and within range of an opponent's intercontinental ballistic missiles (ICBMs). [RAND]

Rapid Repair. The capability to quickly return an airfield to a condition in which it can continue air operations following an attack. [RAND]

- *Level 1:* Conventional civil engineering (CCE) methods, such as aggregate crater fill and laying replacement concrete. [RAND]
- *Level 2:* CCE plus the use of folded fiberglass (FFM) or aluminum mats (e.g., AM-2). [RAND]
- *Level 3:* CCE, FFMs, and Critical Runway Assessment and Repair (CRATR) teams. [RAND]

Recovery. In air operations, that phase of a mission that involves the return of an aircraft to a land base or platform afloat. [JP 1-02, 2014] Also see "Type H: Recovery and Reconstitution" in "Passive Defenses."

Salvo. A method of fire or delivery in which multiple missiles are launched at their targets simultaneously. [JP 1-02, 2014, RAND modified]

For operational resilience, it is useful to consider three levels of salvo:

- *Level 1:* **Small.** Five or fewer missiles. [RAND]
- *Level 2:* **Medium.** 6-20 missiles. [RAND]
- *Level 3:* **Large.** More than 20 missiles. [RAND]

Security Dilemma. The tendency of a nation's policymakers to see any military defensive measures undertaken by an adversary as an act of aggression. [Jervis, *Perception and Misperception in International Politics*, 1976, RAND modified]

Sortie. An operational flight by one aircraft. This can be qualified as a combat sortie, tanker sortie, etc. For analysis of operational resilience, it is important to remember that sorties should only be counted if they contribute to combat power and the achievement of campaign objectives. An aircraft launched from a base does not provide an effective sortie if it lacks the range to reach a target, patrol location, or refueling rendezvous with sufficient armament or loiter time to complete the required mission and return to a recovery base. [JP 3-30, 2014, RAND modified]

Standoff Distance. A distance within the range of friendly weapons or sensors, intended to minimize or eliminate the threat from hostile weapon systems. [RAND]

Temporary Ramp. A metal-mat aircraft parking area that is capable of being relocated. Temporary ramps cannot support long-term operations without periodic repair or replacement. [RAND]

Tent City. Nonhardened area of a military base designated as temporary living quarters for personnel. [RAND]

Turnaround Time. The portion of time during which an aircraft is grounded between sorties that is devoted to takeoff, landing, safety checks, maintenance, and procedural necessities. [JP 1-02, 2014, RAND modified]

Abbreviations

A2/AD	anti-access / area denial
AAA	air defense artillery
AAFIF	Automated Air Facilities Intelligence File
AFDD	Air Force Doctrine Document
ALCM	air-launched cruise missile
BMD	ballistic missile defense
C2	command and control
CAP	combat air patrol
CBRN	chemical, biological, radiological, nuclear
CC&D	camouflage, concealment, and deception
CCE	conventional civil engineering
CEP	circular error probable
CODE	Combat Operations in Denied Environments
CSG	carrier strike group
CTA	central terminal area
DCA	defensive counterair
DoD	U.S. Department of Defense
FFM	folded fiber mats
FOB	forward operating base
FOL	forward operating location
GLCM	ground-launched cruise missile
gpm	gallons per minute
IADS	integrated air defense system
IATA	International Air Transport Association
IRBM	intermediate range ballistic missile
ISR	intelligence, surveillance, and reconnaissance
LOC	line of communication
MOB	main operating base
MOE	measure of effectiveness
MOS	minimum operating surface
MRBM	medium range ballistic missile
ODS	Operation Desert Storm
OEF	Operation Enduring Freedom
OIF	Operation Iraqi Freedom
ORAM	Operational Resilience Analysis Model

PAF	RAND Project AIR FORCE
PACAF	Pacific Air Forces
PGM	precision guided munition
POL	petroleum, oil and lubricants
SAM	surface-to-air missile
SOAR	Strategic and Operational Aspects of Resilience
SRBM	short-range ballistic missile
TBM	tactical ballistic missile

Bibliography

Air Force Doctrine Document 3-10, *Force Protection*, Washington, D.C.: Department of the Air Force, July 28, 2011.

Air Force Instruction 11-202, Volume 3, *General Flight Rules*, Washington, D.C.: Department of the Air Force, October 22, 2010.

Air Force Manual 10-1602, *Nuclear, Biological, Chemical, and Conventional (NBCC) Defense Operations and Standards*, Washington, D.C.: Department of the Air Force, November 2011.

Air Force Manual 10-2503, *Operations in a Chemical, Biological, Radiological, Nuclear, And High-Yield Explosive (CBRNE) Environment*, Washington, D.C.: Department of the Air Force, July 6, 2011.

Air Force Pamphlet 10-1403, *Air Mobility Planning Factors*, Washington, D.C.: Department of the Air Force, December 12, 2011.

British Broadcasting Corporation, "Who, What, Why? How Can an Airport Run Out of Fuel?" *BBC News Magazine,* June 7, 2012. As of October 27, 2015: http://www.bbc.com/news/magazine-18355592

Center for Strategic and Budgetary Assessments, *AirSea Battle: A Point-of-Departure Operational Concept*, Washington, D.C., 2010.

Chankij, Michael K., *Assessing Resilience of the Jet Propellant-8 (JP-8) Distribution System on Guam*, Naval Postgraduate School: Monterey, Calif., June 2012.

Cliff, Roger, *Anti-Access Measures in Chinese Defense Strategy*, testimony presented before the U.S. China Economic and Security Review Commission, January 27, 2011. As of October 29, 2015:
http://www.rand.org/pubs/testimonies/CT354.html

Davis, Paul K., and James H. Bigelow, *Experiments in Multiresolution Modeling (MRM)*, Santa Monica, Calif., RAND Corporation, MR-1004-DARPA, 1998. As of October 27, 2015: http://www.rand.org/pubs/monograph_reports/MR1004.html

Davis, Paul K., and Paul Dreyer, *RAND's Portfolio Analysis Tool (PAT): Theory, Methods and Reference Manual*, Santa Monica, Calif.: RAND Corporation, TR-756-OSD, 2009. As of October 27, 2015:
http://www.rand.org/pubs/technical_reports/TR756.html

Davis, Paul K., R. D. Shaver, and Justin Beck, *Portfolio-Analysis Methods for Assessing Capability Options*, Santa Monica, Calif.: RAND Corporation, MG-662-OSD, 2008. As of

October 27, 2015:
http://www.rand.org/pubs/monographs/MG662.html

DoD—*See* U.S. Department of Defense.

Emerson, Donald E., *TSAR and TSARINA: Simulation Models for Assessing Force Generation and Logistics Support in a Combat Environment*, Santa Monica, Calif.: RAND Corporation, P-6773, 1982. As of October 27, 2015:
http://www.rand.org/pubs/papers/P6773.html

Hagen, Jeff, Patrick Mills, and Stephen M. Worman, *Analysis of Air Operations from Basing in Northern Australia*, Santa Monica, Calif.: RAND Corporation, TR-1306-AF, March 2013, not available to the general public.

IATA—*See* International Air Transport Association.

International Air Transport Association, *IATA Guidance on Airport Fuel Storage Capacity EDITION 1*, May 2008. As of October 27, 2015:
https://www.iata.org/policy/Documents/guidance-fuel-storage-may08.pdf

Jane's Information Group, *Jane's All the World's Aircraft*, Englewood, Colo.: IHS Jane's, 2013.

Jervis, Robert, *Perception and Misperception in International Politics*, Princeton, N.J.: Princeton University Press, 1976.

JOAC—*See* Joint Operational Access Concept.

Joint Operational Access Concept, version 1.0, Washington, D.C. : U.S. Department of Defense, January 17, 2012.

Joint Publication 1-02, *Department of Defense Dictionary of Military and Associated Terms*, Washington, D.C.: Office of the Joint Chiefs of Staff, November 8, 2010, as amended through March 15, 2014.

Joint Publication 3-01, *Countering Air and Missile Threats*, Washington, D.C.: Office of the Joint Chiefs of Staff, March 23, 2012.

Joint Publication 3-03, *Joint Interdiction*, Washington, D.C.: Office of the Joint Chiefs of Staff, October 14, 2011.

Joint Publication 3-11, *Countering Air and Missile Threats*, Washington, D.C.: Office of the Joint Chiefs of Staff, *Operations in Chemical, Biological, Radiological, and Nuclear Environments*, Washington, D.C.: Office of the Joint Chiefs of Staff, October 4, 2013.

Joint Publication 3-15.1, *Counter–Improvised Explosive Device Operations*, D.C.: Office of the Joint Chiefs of Staff, January 9, 2012.

Joint Publication 3-17, *Air Mobility Operations*, Washington, D.C.: Office of the Joint Chiefs of Staff, September 30, 2013.

Joint Publication 3-30, *Command and Control of Joint Air Operations*, Washington, D.C.: Office of the Joint Chiefs of Staff, February 10, 2014.

Joint Publication 3-35, *Deployment and Redeployment Operations*, Washington, D.C.: Office of the Joint Chiefs of Staff, January 31, 2013.

JP—*See* Joint Publication.

Krepinevich, Andrew, Barry Watts, and Robert Work, *Meeting the Anti-Access and Area-Denial Challenge*, Washington, D.C.: Center for Strategic and Budgetary Assessment, 2003. As of October 27, 2015: http://www.csbaonline.org/publications/2003/05/a2ad-anti-access-area-denial/

Markowitz, Harry, "Portfolio Selection," *Journal of Finance*, Vol. 7, No. 1., pp. 77–91, March 1952.

Mellerski, R. Craig, and Craig Rutland, "The New Face of Rapid Runway Repair," *Air Force Civil Engineer*, Vol. 17, No. 3, 2009. As of October 27, 2015: http://www.afcec.af.mil/shared/media/document/AFD-120929-010.pdf

Moynihan Richard A., *Investment Analysis using the Portfolio Analysis Machine (PALMA) Tool*, MITRE Corp., 21 July 2005. As of October 27, 2015: http://www.mitre.org/sites/default/files/pdf/05_0848.pdf

National Geospatial-Intelligence Agency, *Automated Air Facilities Intelligence File (AAFIF)*, St. Louis, Mo., 2010, not available to the general public.

Office of the Secretary of Defense, Annual Report to Congress, *Military and Security Developments Involving the People's Republic of China 2013*, Washington, D.C., 2013.

Pettyjohn, Stacie L., and Alan J. Vick, *The Posture Triangle: A New Framework for U.S. Air Force Global Presence,* Santa Monica, Calif.: RAND Corporation, RR-402-AF, 2013. As of October 27, 2015: http://www.rand.org/pubs/research_reports/RR402.html

Przemieniecki, J. S., *Mathematical Methods in Defense Analyses, Third Edition,* Reston, Va.: American Institute of Aeronautics and Astronautics, 2000.

Schneider, Barry R., *Counterforce Targeting Capabilities and Challenges*, Maxwell Air Force Base, Ala.: Air University, August 2004.

Shlapak, David A., David T. Orletsky, Toy I. Reid, Murray Scot Tanner, and Barry Wilson, *A Question of Balance: Political Context and Military Aspects of the China-Taiwan Dispute*,

Santa Monica, Calif.: RAND Corporation, MG-888-SRF, 2009. As of October 27, 2015: http://www.rand.org/pubs/monographs/MG888.html

Stillion, John, and David T. Orletsky, *Airbase Vulnerability to Conventional Cruise missile and Ballistic-Missile Attacks: Technology, Scenarios, and U.S. Air Force Responses,* Santa Monica, Calif.: RAND Corporation, MR-1028-AF, 1999. As of October 27, 2015: http://www.rand.org/pubs/monograph_reports/MR1028.html

Thomas, Brent, Mahyar A. Amouzegar, Rachel Costello, Robert A. Guffey, Andrew Karode, Christopher Lynch, Kristin F. Lynch, Ken Munson, Chad J. R. Ohlandt, Daniel M. Romano, Ricardo Sanchez, Robert S. Tripp, and Joseph V. Vesely, *Project AIR FORCE Modeling Capabilities for Support of Combat Operations in Denied Environments*, Santa Monica, Calif.: RAND Corporation, RR-427-AF, 2015. As of October 27, 2015: http://www.rand.org/pubs/research_reports/RR427.html

U.S. Army and U.S. Marine Corps, *Gaining and Maintaining Access: An Army–Marine Corps Concept*, Washington, D.C., March 2012.

U.S. Department of Defense, *Unified Facilities Criteria (UFC): Operation and Maintenance: Maintenance of Petroleum Systems*, Washington, D.C., UFC 3-460-03, January 21. 2003. As of August 3, 2013:
http://www.wbdg.org/ccb/DOD/UFC/ufc_3_460_03.pdf

Vick, Alan J., and Jacob L. Heim, *Assessing U.S. Air Force Basing Options in East Asia, Santa Monica, Calif.: RAND Corporation*, MG-1204-AF, 2013, not available to the general public.